# Advances in Lectin Research     Volume 1

W0235095

Advances in Lectin Research          Volume 1

# Advances in Lectin Research

## Volume 1

*Edited by Hartmut Franz*

*Coeditors: Ken-ichi Kasai, Jan Kocourek,*
          *Sjur Olsnes, Leland M. Shannon*

*Preface by Irwin J. Goldstein*

*With 44 Figures*

Springer-Verlag Berlin Heidelberg GmbH

Hartmut Franz
Staatliches Institut für Immunpräparate und Nährmedien
Klement-Gottwald-Allee 317–321, Berlin, DDR-1120

Ken-ichi Kasai
Faculty of Pharmaceutical Sciences, Teikyo University
Sagamiko, Kanagawa 199-01, Japan

Jan Kocourek
Dept. of Biochemistry Charles University Prague
Hrusicka 2515, 14100 Praha 4 – Spořilov 11, Czechoslovakia

Sjur Olsnes
Norsk Hydrol's Institute for Cancer Research, Montebello Oslo 3, Norway

Leland M. Shannon
University of California, Riverside, California 92521, USA

Irwin J. Goldstein
Dept. of Biological Chemistry University of Michigan
Ann Arbor, MI 48109, USA

Sole distribution rights for all non-socialist countries
Springer-Verlag Berlin Heidelberg GmbH.

ISBN 978-3-662-11059-1     ISBN 978-3-662-11057-7 (eBook)
DOI 10.1007/978-3-662-11057-7

2131/3140-543210

# Authors

*Dr. Christa Beurton*
Humboldt-Universität zu Berlin, Bereich Botanik,
Arboretum des Museums für Naturkunde
Späthstraße 80–81, Berlin DDR-1195

*Dr. Edilbert van Driessche*
Instituut voor Moleculaire Biologie,
Vrije Universiteit Brussel
Paardenstraat 65, B 1640 St. Genesius-Rode, Belgium

*Prof. Dr. Dr. Hartmut Franz*
Staatliches Institut für Immunpräparate und Nährmedien
Klement-Gottwald-Allee 317–321, Berlin, DDR-1120

*Renate Israel*
Greifswalder Straße 13, Berlin, DDR-1055

*Prof. Dr. H. Rüdiger*
Institut für Pharmazie und Lebensmittelchemie
der Universität Würzburg
Am Hubland, Würzburg, D-8700

# Preface

Progress in the field of lectin chemistry and biochemistry has been truly astounding during the past decade. New lectins of unique carbohydrate binding specificity have been discovered; the hydrophobic sites of lectins have been probed; lectin genes have been cloned and translated, opening the way for site directed mutagenesis and the identification of amino acid residues essential for the biological activity of lectins and, even more exciting, the possibility of designing lectin binding sites of defined specificity. Libraries of the amino acid sequences of lectins from several plant families have been assembled making possible the construction of phylogenetic trees and providing new insights into the evolutionary development of plant species. The X-ray crystallographic structures of several lectins have been solved (e. g. concanavalin A, wheat germ agglutinin, *Vicia faba* and pea) and many more are "in the works", e. g. soybean and peanut agglutinins, providing us with an in-depth view of the molecular architecture and topography of this truly fascinating group of carbohydrate binding proteins.

What does the future hold for lectin research? We may expect an acceleration of research on the molecular biology of lectins – incorporation and expression of lectin genes in other organisms, and the tayloring of sites which bind specific ligands. An increased application of lectins in biotechnology and biomedical research will surely follow: immobilized lectin columns will be employed for the isolation and resolution of cell populations and subcellular organelles; the isolation of medically-important glycoproteins (e. g. immunoglobulins, enzymes, blood coagulation factors); and as an analytical tool for the characterization of glycoconjugates. Lectin will also be employed to target drugs to specific cells and tissues, and for the diagnosis of various disease states. We may also expect the discovery of additional lectin-like molecules in animal species. Finally, there will be increased activity in the important areas of the metabolism and physiological role of plant and animal lectins.

The present series, *Advances in Lectin Research*, will address the need to present accounts of the most recent advances in all areas of lectin research. It is a timely series which should be greeted with widespread acceptance and enthusiasm.

*Irwin J. Goldstein, Ann Arbor*

# Foreword

The years 1987/88 appear to be a suitable moment for the publication of *Advances in Lectin Research*. The detection of ricin 100 years ago led to interesting results concerning sugar-binding proteins, now called lectins. Today two main lines exist in "lectinology". Firstly, lectins are used as multi-purpose tools (analysis and isolation of glycoconjugates, characterization and preparation of cells and microorganism, lectin histochemistry). Secondly, lectins are of interest concerning their physiological roles and other biological activities. Especially during the last 10 years, the latter aspects have been widely noticed. More and more the opinion is being accepted that lectins are glues of moderate affinity, making the contact of biologically cooperating glycoconjugates and glycoconjugates bearing cells possible.

The results of lectin research anticipated for the coming years should be of great importance not only for basic research, but also for diagnostic and even therapeutic application (adherence inhibition of bacteria and tumor cells, immunmodulation, lectin-mediated cytotoxicity, chemically modified toxic lectins). Although monoclonal antibodies may gradually displace lectins as tools, the relation between lectins and antibodies represents an exciting problem. The editor proposed (1981) a phylogenetic connection between lectins and antibodies. Some recent results seem to support this hypothesis. It is the aim of our series to reflect the further development of lectin research. *Advances in Lectin Research*, as a collection of review articles, is not in competition with textbooks and congress proceedings. Included are chapters of common interest such as *Illustrations of Lectin-Producing Plants*.

Editor and coeditors hope that *Advances in Lectin Research* will have a friendly reception. Interested colleagues are invited to cooperate in order to make subsequent volumes as useful and effective as possible.

The editor wants to express his gratitude to the authors and to the coeditors. He also wishes to thank the lector of the "Verlag Volk und Gesundheit", Mrs. Hintz for the fruitful cooperation and Dr. Peter Ziska for helpful suggestions. He is greatly indebted to Dr. Rolf Wachowius for his constant support and thanks also Dr. Michael Gelbin and Dipl. Chem. Heinz Zorn for reviewing the manuscripts. Mrs. Marianne Lobstein is kindly acknowledged for her correct and patient typing and retyping the papers.

*Hartmut Franz, Berlin*

# Contents

# 1 The Ricin Story

Hartmut Franz

## 1.1 Introduction

There is an extraordinarily comprehensive literature about ricin (surveys for example Lugnier 1974; Olsnes and Pihl 1978).

By placing "The Ricin Story" first in "Advances in Lectin Research" it is not intended that this should be a ricin monograph in the usual sense. Rather it is a representation and appreciation of the ricin preparation by Stillmark and Kobert on the occasion of its one hundredth anniversary. This event had an avalanche-like effect on development and also represents the very beginning of lectinology. One may perhaps say that it initiated the archaic phase of the lectin research. We shall see that this discovery, too, had a previous history and that there were parallel developments. To keep this chapter within certain limits, it will be focused on the following aspects: previous history

Stillmark's dissertation

its enlarged version one year later

direct utilization of its results by Paul Ehrlich

Kobert's retrospective view at this work 25 years later.

In conclusion, there will be a short review in the light of our present knowledge (Scheme 1.1). When looking through older works, the author came across a deeply impressive mosaic consisting of joy of discovery, suffering, errors, ingenuity, accidents and, concerning certain researchers, also researcher's tragedy. He hopes that he has comprehended these connections correctly and can present them in an adequate manner.

## 1.2 Previous History

The seeds of *Ricinus communis L.* and preparation of them have been used for several thousand years. It has been known for a long time that the purgative effect is connected with the castor oil which is obtained from them by pressing. The toxic component of ricinus seeds, however, remained unexplored for a long time. It is only in the second half of the nineteenth century that interest was awakened by aqueous extracts, in which it was assumed, and rightly so, that a toxically acting substance was present.

In 1854 Henry Bower stated: "From this circumstance, and the fact of the beans forming an emulsion with cold water, it was inferred that they contained a peculiar albuminous principle analogous to that existing in almonds; further experiment verified this. A product resembling emulsin was obtained by making an emulsion of the beans with water, adding to this twice its bulk of ether, and after frequent agitation allowing it to stand, when a transparent fluid separated at the bottom of the vessel; this was separated from the supernatant ethereal mixture, and

*Scheme 1.1*   Current data on ricin

At the present time ricin denotes a lectin (molecular weight about 65,000) which consists of an A-chain (molecular weight about 30,000) and a B-chain (molecular weight about 33,000). Both chains are bound by one disulfide bridge. The A-chain, as the toxophoric part of the ricin molecule, inhibits enzymatically the protein synthesis on ribosomal level. The B-chain is the very lectin (haptophoric part). Ricin is highly toxic and agglutinates human erythrocytes only weakly. Two species of ricin have been described (ricin A and ricin B). Each contains a different B-chain. Moreover, two types of A-chains ($A_1$ and $A_2$) are known (Vidal et al. 1985). Another name of ricin is $RCA_{II}$. The amino acid sequence has been published by Funatsu et al. (1979).

The "ricin" of Stillmark contained besides the toxic compound a second lectin (molecular weight about 120,000) which is named ricinus agglutinin or $RCA_I$. It acts strongly agglutinating and is much less toxic than $RCA_{II}$. The ricinus agglutinin consists of two A- and two B-chains which are partially different from the A- and B-chains of ricin. Olsnes and Pihl (1982) published the following scheme demonstrating the molecular structure of the two main lectins of *Ricinus communis* L. As shown recently the A-chain of ricin acts as a N-glycosidase [see Endo Y. (1988) The mechanism of action of ricin and related toxic lectins on the inactivation of eukaryotic ribosomes. In: Advances in Lectin Research, Vol. 2].

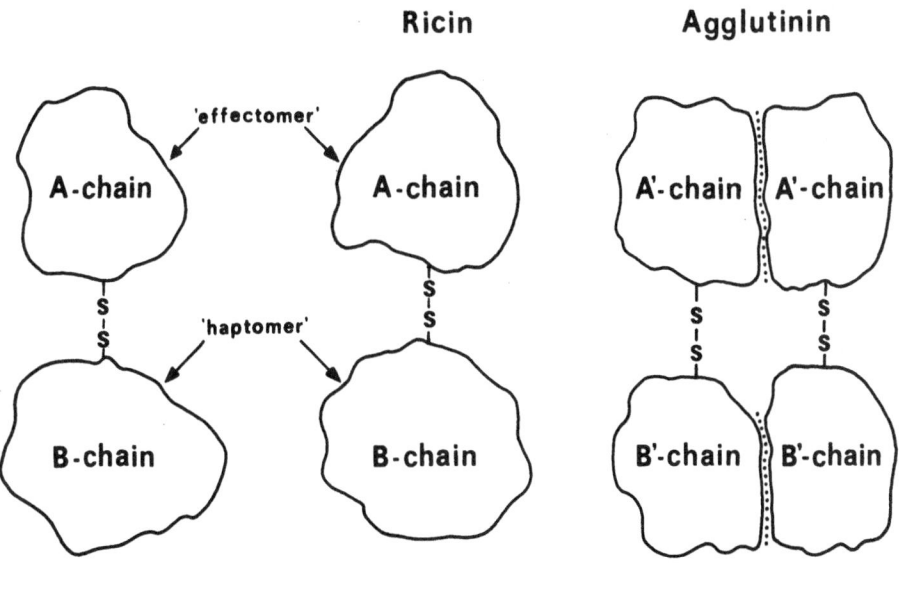

alcohol added, which threw down a precipitate white and flocculent; this was collected upon a filter, and washed with fresh portions of alcohol, and dried under the receiver of an air-pump. This product was soluble in water, the solution reddening litmus paper, and when heated to 212 °C coagulated."

Bower considered the amygdalin-splitting effect to be the most important one of this product which most likely, contained the ricin. However, he did not carry out any investigations with respect to its toxicity. A work by Emil Werner, which appeared in 1870, deserves particular attention. He wrote: "When comparing all experiments that have just been described, I arrived at the conclusion that the best solvent for the active substance of ricinus seeds is water and that this substance is destroyed by a longer lasting contact with hot water. Since I know that the active substance is insoluble in strong alcohol, I decided to precipitate it from an aqueous solution by means of alcohol. So one ounce of fresh seed was crushed with a little water and allowed to stand for 24 hours, agitating it quite frequently. Subsequently it was filtered and mixed with 4 vol. 90% alcohol, the precipitate that had formed was collected on a filter, then spread and dried at normal temperature. The whole quantity of the precipitate given to the dog in one portion (orally?, H.F.) remained without any effect."

As the isolation of the toxic principle does not deviate considerably from later successful methods (see below), it is at first sight amazing that Werner did not find any toxic effect. Dixson (1887), who obviously was also astonished, gave an explanation which, however, certainly is not or only partly acceptable (see below). The failure of the toxicity test should rather be explained by a remark made by Werner elsewhere in the same work: "I should mention here that in all experiments conducted in dogs no drug was administered before the animal was completely restored after the preceding test."

Werner obviously produced a resistance to ricin in his experimental animals 20 years before Paul Ehrlich without knowing or being able to know this. So he had to arrive inevitably at misleading results and interpretations of his experiments. That is why the phenomenon that sublethal ricin doses can produce an immunity against ricin in animals has prevented Emil Werner from being considered today as the discoverer of ricin. On the other hand Paul Ehrlich who ingeniously recognized or foresaw these connections 25 years later, is rightly considered to be one of the founders of immunology.

Proteins (so-called crystallized albumins) were also prepared from ricin press cake by Ritthausen (1882). In 1887, an informative work in two parts was published by Dixson who continued after Bubnow's early death his work at the Schmiedeberg laboratory in Strasbourg. In 1883 shortly before his death Bubnow had obviously extracted ricinus seeds with diluted hydrochloric acid and isolated a toxic substance from these solutions by an addition of alkali. Dixson in particular pointed out that Bubnow had for the first time obtained a toxic substance in a solid form. Endeavoring to improve Bubnow's method he proceeded as follows: "It has been seen that moderately strong acid yielded an inert body in great amount, and a very weak acid yielded nothing, and yet the residue was still active if the acid were not too strong; hence it was probable that the principle was dissolved in the water, and that the acid was merely a solvent for albumen which the alkali precipitated, the precipitate carrying down some of the dissolved acting principle with it, and that if the acid was strong it made plenty of acid albumen (but killed the active body) and so yielded plenty of precipitate (inert 'extract'), and that if weak it made little acid albumen, and gave no precipitate at all. But as the meal was quite unsuitable for making solutions from, as they filtered badly, some castor cake was ordered from Italy ...

This castor cake is merely the residue of the wholecrushed seed after the expression of the oil for commerce. It is exceedingly energetic in action ... I took this 'cake' and rubbed it up with an equal bulk of water, let it stand a few hours, and finally squeezed out the fluid in a press. The resulting liquid was filtered carefully; but, as it did this very slowly and imperfectly, it was found best to let it settle over night, and then filter the yellow supernatant syrupy fluid through filter paper. The result was a clear yellow fluid of great energy. On adding five times as much, 98% alcohol ('absolute alcohol') was all precipitated. Hence we can see why Werner failed to find this body; for it is not precipitated in a large amount of weak alcohol unless present in large amount itself. But this precipitate is always very energetic. Like Bubnow's, it is very full of albumen; it also gives the reaction of a glycoside; when dry it forms a horny yellow mass. The question now was how to purify it. In this all attempts failed, but certain interesting facts were brought out. This precipitate never would retain its activity after heating in water for one instant at the boiling point, and copious precipitate of albumen formed. The seeds when merely heated in alcohol, even to its boiling point, were rendered inert . . . ."

Repeated precipitation with alcohol certainly took more and more yellow coloring matter into the alcohol but, unfortunately, the precipitate previously soluble in water became each time partly insoluble in water, till at last only a trifle of a white powder, soluble in water, but inert, was left, the rest remaining on the filter as a partly active, sticky, albuminous body. Filtration was always very slow, and if the time occupied extended over two or three days the fluid putrefied."

But Dixson arrives at the conclusion that it is not the protein component that is responsible for the toxicity ["... that the active principle is probably the very glycoside found in the seed, which, through a slight change in chemical composition (hydration perhaps), loses its activity"].

More detailed statements regarding the toxicity of the isolated substance are lacking in Dixson's publications. He furthermore proposed to use large amounts of the produced press cake for human nutrition: "... while the cake left behind would, if cooked, form an albuminous, pleasant article of food, especially of use in India, where thousands of tons of the cake are made yearly in a crude way. In famine, this would be a valuable adjunct to rice, seeing that meat is not eaten by Hindoos."

## 1.3    The Investigations of Stillmark and Kobert

On 17 February 1888[1] an inaugural dissertation dealing with the subject *Ricin, a toxic enzyme from the seeds of Ricinus communis L. and some other Euphorbiaceae* (Fig. 1.1) was given the printing licence of the Medical Faculty of the Dorpat university in Estonia (now Tartu, Estonian Soviet Republic, USSR). The author, Hermann Stillmark, performed these investigations at the suggestion and under the direct guidance of the director of his institute Prof. Dr. Rudolf Kobert (Stillmark 1888). Kobert (Fig. 1.2), who was born in Bitterfeld in 1854, was admitted as an orphan to the famous Francke's Foundation in Halle und was later given the possibility to study medicine at the university of that town. From 1880 to 1886 he worked at Schmiedeberg's laboratory in Strasbourg and then accepted a call to Dorpat to work as director of the pharmacological institute. In 1897 he had to leave Dorpat, was in charge of the Görbersdorf tuberculosis hospital in Silesia for 2 years and in 1899 received a call to Rostock to take over the direction of the Institute of Pharmacology and Physiological Chemistry. On 27th December 1918 he met with sudden death on his way to the institute (Rosbach 1969).

Peter Hermann Stillmark (Fig. 1.3) was born in Pensa on the 22 July 1860. He attended the "Ritter- und Domschule" in Reval (now Tallinn, Estonian Soviet Republic, USSR) and subsequently studied medicine in Dorpat. After his graduation he left the university and from 1888 worked as a physician in Merjama and later in Helmet (both in Estonia). From 1895 up to his death on June 23 1923, he was a factory, railway, school and prison doctor in Pernau (Estonia).

Because of the above mentioned great importance of Stillmark's work and the interest it still deserves today, a brief survey of its contents will now be given. In his introduction the author gives a historical general account of the therapeutic applications of the ricinus seeds and draws the conclusion that in antiquity only its purgative effect was known (castor oil), whereas in recent times also a toxic activity of both the seeds and the press cake has received attention. As a basic

---

1    Later Kobert wrote that Stillmark's dissertation already appeared in 1887 (Stillmark 1889). However, I did not find any clues for this.

# Ueber

# Ricin, ein giftiges Ferment

## aus den Samen von Ricinus comm. L. und einigen anderen Euphorbiaceen.

### Inaugural-Dissertation

zur Erlangung des Grades

eines

## Doctors der Medicin

verfasst und mit Bewilligung

Einer Hochverordneten Medicinischen Facultät der Kaiserlichen Universität

### zu Dorpat

### zur öffentlichen Vertheidigung bestimmt

von

### Hermann Stillmark,

Estonus.

Ordentliche Opponenten

Prof. Dr. B. Körber. — Prof. Dr. M. Runge. — Prof. Dr. R. Kobert.

### Dorpat.

Schnakenburg's Buchdruckerei.

1888.

*Fig. 1.1*    Title-page of Stillmark's thesis

*Fig. 1.2* Rudolf Kobert (photograph from H.J. Rosbach, Rostock)

*Fig. 1.3* Peter Hermann Stillmark (The author thanks Mrs. B. Callies for her kindly providing the photograph)

concern of his work he formulated: "The purpose of this work is to obtain a more exact knowledge of the toxic principle of the ricinus seeds that has not been investigated thoroughly enough until now." Then a report is given on the experiments carried out until then for a chemical characterization of the ingredients of *Ricinus communis*, dealing also in greater detail with the already mentioned work by Bower, Ritthausen and Werner. Dixson's studies, however, were not known to Stillmark as can be seen from an enlarged edition of his dissertation published in 1889 (Stillmark 1889). There it reads: "... so another work should be mentioned which, after this dissertation had been printed, because available in the form of an abstract (the copy of the original work sent by the author Dixson to Prof. Kobert[1] in Dorpat obviously got lost, and the journal in which the work was published is not available in Dorpat. So all I could do was to reproduce a lecture given here by Husemann." This paper had obviously been incomplete, so that Stillmark's rendering of the text was not correct in all points. This has already been pointed out by Cushny (1898).

In his account of his own studies Stillmark speaks of "an organic component" called by him "ricin for reasons of convenience". He says: "... I have used the way of pharmacological isolation, i. e., a combination of chemical operations, constantly carrying out control experiments regarding the effect in the animal and in the blood, and in this way I finally separated an extremely poisonous substance from the seeds." Stillmark used ricinus press cake, fresh and at least 30-year-old ricinus seeds. The latter were husked and, like the press cake, crushed with a mortar and extracted. The following extracting agents were employed:

– Distilled water; precipitation with soda or lead acetate (both methods were soon rejected) or with acetic acid and $K_4[Fe(CN)_6]$.
– Extractions with 10 % NaCl, precipitation
  • by acetic acid (small yield)
  • in accordance with Sidney Martin precipitation with magnesium and sodium sulfate.

1 It is very probable that Kobert got to know Bubnow's works, and maybe Dixson's works, during his stay in Strasbourg. The tradition of the Strasbourg ricin research has recently been continued by Dirheimer and Lugnier (Lugnier 1974).

The latter method was to serve for the separation of plant albumin substances. Other methods (alkaline extraction after Ritthausen (1882) and precipitation with acetic acid as well as acid extraction and precipitation with $K_4[Fe(CN)_6]$-glycerol extraction were only employed occasionally. The original text of the method of preparation preferred by Stillmark (1888) reads as follows: "As the reader will have noticed, any heating should be avoided in the above described preparations. This is not accidental but the result of many experimental failures which are not communicated here. The reason is that when boiling its solution ricin loses its effect almost instantly. Under such conditions it was difficult to prepare the pure substance in a dry form. The simplest method is as follows: Shelled fresh ricinus seeds which had been thoroughly pressed at a pressure of 30 atmospheres are pulverized and extracted with a 10 % NaCl solution. The filtered waterclear extract is saturated at the same time with magnesium sulphate and sodium sulphate at room temperature and kept in a cold place, giving rise to inch-long crystals of the two sulphates and moreover to a precipitate of white colour, which can readily be separated from these crystals. This precipitate is filtered cold and placed unwashed in a dialysing tube of parchment paper, then hung in ordinary flowing water for three days and subsequently for another three days in distilled (frequently changed) water. The contents of the tube, which are adhering firmly to the wall, is scraped off several times a day, and if the pores get obstructed the contents should be again and again transferred to a new tube. The temperature should be kept low, so that the occurrence of putrefaction is prevented. The beginning of disintegration can be recognized by the occurrence of a cheese-like odour. At the end of the dialysing process the smeary contents of the dialysing tube are scraped off and dried in vacuum over sulphuric acid to obtain an amorphous crust. After complete drying, the crusts can easily be reduced to an odourless, snow-white powder, which by the way in none of the many experiments has been found to be completely free from ashes. It rather seems that a certain amount of the sulphates of 10 to 20 % is retained in the dialysis. But in view of the extremely small doses required for the poisoning of animals this ash content does not bring about the least disturbance. When calculating the dose, this was of course always deducted. By reprecipitating this ash content can be reduced still further. When adhering to this prescription even unskilled workers will succeed in preparing the ricin immediately. If an odour occurs in the dialysis, this does not mean that the preparation is absolutely useless since it is possible to prepare odourless, active ricin from smelling masses."

In this way he received, related to air-dried, shelled seeds, a ricin yield of 2.8 %. His chemical investigations concerning the nature of ricin are summed up by Stillmark as follows: "Ricin is a protein, possibly a phytalbumose and neither an alkaloid nor a glycoside or an organic acid." Contrary to the information of Bower (1854) and Werner (1870), ricin does not release hydrocyanic acid. On the other hand, Stillmark tends to consider ricin to belong to the unformed enzymes (unlike formed substances, e. g. bacteria). In his opinion, criteria of the enzyme nature are water solubility, relative temperature resistance in a solid condition, temperature lability in aqueous solution, precipitability without a greater inactivation by alcohol, solubility in glycerol and an assumed influence on the fibrin coagulation of blood.

"When comparing these characteristics of the unformed enzymes with the properties shown by the studied bodies, the idea, even the conviction suggests itself that this is a genuine enzyme or at least a substance very similar to enzymes. On the other hand we must not deny that a chemist will not see any conclusive proof."

The pharmacological part begins with the description of the effect of ricin on the blood. It is introduced with the following sentence: "The by far most interesting effect of ricin is that exerted on the blood." A basic experiment carried out by Stillmark was the agglutination of (defibrinated) rabbit blood (1 ml in 18 ml physiological sodium chloride solution) by 1 ml ricin solution (1 mg/ml physiological saline solution). When using ricin-free blood samples, the filtration

10 min later showed almost all erythrocytes in the filtrate whereas after a treatment with ricin the erythrocytes remained on the filter. This basic experiment is recommended by Stillmark for the separation of the erythrocytes from the blood serum. He furthermore observed a different agglutination behavior of blood corpuscles in different animal species. He also found an agglutination when ricin was acting on washed erythrocytes of horses.

This process is not specially interpreted by Stillmark, but he obviously is of the opinion that in the presence of ricin in the serum a fibrin-like substance is formed due to the enzymatic activity of ricin. He carried out flow experiments in bovine kidneys which showed a pronounced reduction of the flow volume in blood to which ricin had been added. The addition of ricin to native blood (28 mg to 21 ml blood) brings about a certain delay of blood coagulation. Even when used in relatively high doses, ricin did not exert any influence on William's frog heart. The same was found by him with respect to the excitability of the isolated frog muscle. He found no influence on the isolated nervus ischiadicus of the frog and observed the same qualitative effect on the entire organism of the experimental animal both after subcutaneous and intravenous application and after oral administration. A particularly striking phenomenon was a major inflammation of the small intestine and to a lesser degree of the large intestine and the stomach, which he considered to be due to the effect of the ricin on erythrocytes. In his opinion the agglutination causes disturbances of the microcirculation. Finally he assumes that also thrombosis of the cerebral vessels is caused by ricin.

Stillmark's work further contains a collection of case histories of ricinus intoxications in man followed by records of experiments regarding intoxications. From a comparison Stillmark draws the conclusion that there is agreement of the more important symptoms and the pathological-anatomical changes. In particular he still points to a well-rounded filling of the urinary bladder and the gallbladder. Stillmark found toxic substances also in other ricinus varieties. In *Croton tiglium* a toxic substance agglutinating erythrocytes could also be proved by exploying the method described for ricin. Seeds from *Jatropha curcas*, too, contained a toxic substance, while the examined seeds of *Hura crepitans* and *Hyaenauche globulosa* were free from it. No information is available about experiments regarding the hemagglutination by *Hura crepitans*. Below you will find Stillmark's conclusion in unabridged form:

"1) The only or at least most important toxic component of the ricinus seeds is 'ricin'.

2) Ricin is a protein and belongs to the so-called unformed enzymes.

3) The action of dry heat (even more than 100 °C) is tolerated by ricin without losing its effectiveness; boiling, however, makes it ineffective.

4) The seeds of other Euphorbiaceae also contain ricin or a similar substance.

5) When administered by mouth, by subcutaneous or intravenous injection ricin exerts in a qualitatively similar manner a typical action on the intestine.

6) In defibrinated serum-containing blood ricin produces agglutination of the red blood corpuscles with the formation of a fibrin-like substance.

7) When applied in large doses ricin delays the coagulation of the freshly taken, non-defibrinated blood.

8) Ricin makes defibrinated blood capable of being filtered, i. e. the clear serum forms the filtrate while the red blood corpuscles remain on the filter.

9) This effect is already produced by ricin when it is used in a very high dilution (1:60,000).

10) The blood of different animal species seems to show a gradually different behavior towards this ricin effect.

11) Ricin does not exert an action on the isolated nerve and on the blood pressure, nor at the most a minimum effect on the isolated muscle and William's frog heart.

17

12) As low a dose as 0.1 milligrammes of ricin per kilo of animal applied by the venous route has a fatal outcome, i. e., a ratio of amount of poison to body-weight of 1:10 million.
13) Ricinus seeds cannot be used therapeutically; nor can the ricinus press cake. A utilization of the press cake for physiological-chemical purposes, however, may well be taken into consideration.
14) The effect of castor oil has nothing to do with that of the ricin."

A reprint of Stillmark's graduation work appeared in 1889 in *Arbeiten aus dem Pharmakologischen Institut Dorpat* (Works from the Pharmacological Institute Dorpat), which was edited by Rudolf Kobert (Stillmark 1889). There is the note "with additions of the editor".
On the whole, the agglutinating effect of ricin is accentuated as can be seen from examples. On page 76 there is a footnote saying that the testing of the toxic effect of ricin in the course of the preparation has for the first time been carried out with blood, not in a living animal. Stillmark states with satisfaction: "This extremely humane testing method with blood has been employed by none of the many ricinus investigators before me." The introductory sentence of the pharmacological part now is "the by far most interesting and *most important* effect of ricin is that on the blood". Further: "Professor Kobert has subjected the experiments with blood to a careful revision and extension and for this has also used pure dry ricin ..." It can be seen that drying diminishes the effect somewhat, but not considerably. On page 90 the agglutination of defibrinated blood is described. It reads as follows: "The just described, so very remarkable process makes one first of all think of the question what parts of the blood it uses, the serum or the blood corpuscles. It furthermore obliges us to examine the relations with the fibrin. Most of the experiments necessary for this were performed by Prof. Kobert himself who handed them to me for publication."
On the same page the addition of ricin to the serum of dogs is described. Whereas Stillmark did not find any changes caused by ricin in his original dissertation, he now says: "... where within five hours a distinct turbidity is produced." This sentence is provided with a footnote: "Herr Stillmark has overlooked this change, but I have never missed it. Kobert." Here there may been the chance of an early understanding of the real lectin effect.
On page 96 the term "ricin fibrin" is ihtroduced for the hemolyzed agglutinate resulting from the action of ricin. "I should like to briefly call this body ricin fibrin, in order to intimate that it shows many a similarity with genuine fibrin but must indeed be considered to be quite different from it."
On page 145 he additionally says with regard to the importance of the ricin agglutination in vivo: "It must suffice to emphasize that Prof. Kobert found in the intestinal mucosa the vascular loops of the villi stuffed with blood corpuscles which were about to conglutinate, even when he did not carry out the experiment after the spontaneous death but in previously slaughtered animals under consideration of all precautions. The supposition which I had at the beginning that all large and small vessels were always stuffed with coarse clots did not prove true. A loose adhesion of the blood corpuscles among each other is already sufficient to produce the most serious circulatory disturbances in the intestine as well as in the brain. Further microscopic work will bring the necessary light. For me it is sufficient to have initiated it." On page 151 Stillmark closes with the words: "At any rate I think I am entitled to demand that those who will pass a depreciatory judgement on my work substantiate it by their own experiments. Prof. Kobert by making ricin commercially available has made it easy enough to conduct such tests." He obviously refers to the fact that Kobert had suggested to the firm of Merck to make ricin prepared according to Stillmark's method commercially available. As early as in October 1899 the price list of Merck contained "Ricin after Professor Kobert"[1]. In the annual report of the firm of Merck of

January 1890 there is an entry concerning ricin that represents a summary of the results of Still-mark's work. As regards the toxicity we find: "Ricin is an extremely poisonous body which has no relation to the well-known effects of castor oil. Applied by the intravenous route, ricin kills already in doses below 0.03 mg/kg of body-weight. Given by mouth, the fatal amount of ricin administered to a person of 60 kg body-weight would be about 18 g. Administered in a lethal dose ricin produces a hemorrhagic gastroenteritis resulting after a rapid decline of the forces in a lethal outcome with convulsions and collapse in a somnolent condition."

It should be particularly pointed out that in the same annual report abrin is also mentioned although the respective dissertation by Hellin, who was also a pupil of Kobert, only appeared in 1891 (Hellin 1891). About abrin we read: "Abrin obtained from the seed of *Abrus precatorius* forms a brownish-yellow, water-soluble powder. It is an extremely poisonous protein, which like ricin belongs to the class of the so-called 'unformed enzymes'. Abrin was prepared by me at the request of Prof. Kobert in Dorpat who at present is subjecting it to a thorough pharmacological examination. According to information given by this research worker in letters the lethal dose per kilogramme of body-weight is 0.00001 g when introduced directly into the blood stream. The enormous toxicity of this body requires the greatest caution both when keeping and when applying it. It is certainly of interest to medical gentlemen to learn that Prof. Kobert was able to produce the so-called jequirity ophthalmin [see also Sidney Martin and Wolfenden: *The Active Principles of the Seeds of Abrus precatorius (jequirity)*, The Brit. Med. Journ. 1880, 1053]."

The intuitive realization of the significance of ricin and abrin by Kobert and an adequate reaction of the firm of Merck have made possible at that time the epochal studies of Paul Ehrlich. They will now be dealt with in greater detail.

## 1.4 Paul Ehrlich's Investigations about Ricin and Abrin

Paul Ehrlich (1891a) (Fig. 1.4) immediately recognized the relations between ricin and abrin on the one hand and certain bacterial toxins on the other. He writes: "Since at the time being the toxalbumoses produced by pathogenic bacteria have not been isolated yet in an absolutely pure condition and, as is easily understandable, larger quantities are for the time being only accessible with difficulty, I have used two poisonous proteins for my experiments, which stem from the plant kingdom, namely ricin, the toxalbumin of ricin seeds, and abrin, the effective principle of the jequirity bean. Owing to the initiative of Kobert, who has deserved well of the physiological research of this substance, both bodies are obtainable in a sufficient purity and in abundance."

Paul Ehrlich "ricinated" white mice, preferably by feeding ricin, and in this way he obtained "ricin-resistent" animals. A ricin immunity occurred in particular after the 6th day. Paul Ehrlich observed that the immunized animals tolerated injections reaching up to 800 times the value of the lethal dose! An immunization effect could also be proved on the eye. For determining the lethal dose, Paul Ehrlich usually injected 1 ml of the ricin solution. In some cases solutions of Merck's ricin (ash content 30 %) proved to be lethal in a dilution of 1:750,000, but always in the dilution of 1:200,000. The animals died after 2 to 4 days. Guinea pigs were considerably less sensitive than white mice.

A short time later the study *Über Abrin* appeared (Ehrlich 1891b). Here he writes: "Having promoted in an excellent manner the recognition of the physiological effect of abrin is again the merit of the versatile Dorpat pharmacological institute. The following statements are taken

1   I am greatly indebted to Dr. Possehl of the firm of Merck for kindly handing over material from the firm's archives.

*Fig. 1.4*  Paul Ehrlich (1854–1915)

from a short lecture by Kobert and the detailed work of his pupil Hellin. So abrin is an extremely toxic albumose, which with respect to its effects shows the greatest similarity to ricin. The basic effect of both consists in the capability of coagulating the blood in a strange manner and in this way produce multiple thrombosis especially of the intestinal vessels. The post-mortem findings are exactly the same, no matter whether it is ricin or abrin. They have furthermore in common that both exert an effect on the mucous membrane of the eyes as well as the property to change over to non-toxic substances under the influence of digestive enzymes such as pepsin and trypsin. Because of this far-reaching correspondence one has taken into consideration at the Dorpat institute for a time the possibility of an identity of the two substances. According to the latest publication by Hellin this viewpoint has now been discarded and one believes at present that the two bodies are extremely similar but not identical. After until now the pharmacological investigations did not show any sharp differences between the two bodies, I must now regard it as a favourable fact that my experiments have yielded quite considerable differences which make it necessary to consider ricin and abrin to be quite different bodies."

Paul Ehrlich arrived at the conclusion that ricin-resistent animals were just as sensitive to abrin as completely normal ones. The same is true of abrin-resistant animals with respect to ricin.

The importance of the two above mentioned articles by Paul Ehrlich to the field of immunology is generally recognized today. This also applies to the production of antitoxic antibodies, to their quantitative control and thus to the development of the foundations of a passive immunization and the serum therapy connected with it, which proved to be extraordinarily beneficial in the first half of our century. It should also be pointed out in this connection that Paul Ehrlich for the first time used immunological methods for the distinction of chemically very similar substances like ricin and abrin.


## 1.5    Kobert's Review of the Year 1913

Under the title *Beiträge zur Kenntnis der vegetabilischen Hämagglutinine* (Contributions to the Knowledge of Plant Hemagglutinins) Kobert once more gave his opinion about the ricin problems in a detailed manner (Kobert 1913). First of all, he points out that he induced Stillmark in 1888 to introduce the designation ricin (his spelling now is "rizin") just as he induced Elfstrand[1] to introduce the term "haemagglutination" in 1898. He then mentions besides the original

Stillmark's technique a second method for the preparation of ricin: "A second method that is much simpler and can be used for all plant agglutinins at least for our purposes consists in the precipitation of the concentrated, 0.9 % sodium chloride containing aqueous extract with the at least equal volume of alcohol, after-washing with little alcohol and drying of the precipitate in a vacuum without heating on clay plates, which have been thoroughly dried and slightly heated before. Choosing quite a large amount of alcohol for precipitating the extract offers the advantage that one can very soon start filtration and that this will proceed well. On the other hand, however, the preparation becomes more and more insoluble in dependence on alcohol concentration and time of interaction, respectively."

Previous authors who also used the alcohol precipitation of aqueous ricinus extracts (Bower, Werner, Dixson), however are not mentioned by Kobert. Lau (1901) mentioned that Kobert in the meantime also proposed to make Stillmark's preparation commercially available in an undialyzed form. The user had to dialyze it himself immediately before use. The stability of the substance should in this way be improved.

Kobert points out that although both methods fail to yield a pure preparation, they furnish one that is sufficient for all biological studies. For obtaining preparations of a higher degree of purity he refers to Osborne who used the fractionated precipitation with ammonium sulfate solutions. Of particular importance are Kobert's remarks about the binding of ricin to different cells: "Rizin has the capability of combining chemically with many lipoid-rich cells of the body of human and animals. I already studied this together with C. Lau in 1900. It regards isolated brain, liver, and kidney cells, etc. I have just seen this again together with Reid." (see Reid 1913, H. F.) In both cases our technique was to suspend to completely bloodfree cells which are separated from coarser particles by a fine network of meshes, in physiological sodium chloride solution and to mix them with rizin dissolved in physiological saline solution. Then one often sees that after only a few minutes the cells agglomerate to form larger flakes while in the control glass nothing of that kind can be observed. Of course one can also perfuse a suitable organ such as the liver with Ringer's solution after having completely removed the blood, something that has for instance been done by W. N. Woronzow. In this process the liver cells anchor the poison and the outflowing liquid finally becomes completely free from poison. In an analogous manner the red blood corpuscles can also anchor the rizin."

Lau (1901) wrote the following: "The isolated cells or cell particles of the individual organs suspended in physiological NaCl solution are agglutinated in a pronounced manner by the four poisons." (Ricin, abrin, crotin and robin, H. F.) "The agglutination can most clearly be seen in case of the isolated liver cells.

... In his experiments Stillmark has probably too much emphasized the effect of ricin on the blood corpuscles ... The agglutinating plant poisons have the capability of changing very different cellular structures of the warm-blooded organism in a strange manner and in this way cause their becoming devitalized (brain, liver, kidneys, red and white blood cells, etc.). All these effects must be taken into account when judging the overall effect of these poisons. For testing the poisons in the laboratory their effect on the blood is most important because this can be carried out in the easiest way and because, if present, conclusions may also be drawn with respect to other effects."

So Lau was the first to make a statement about the cytotoxicity of ricin and the other substances reacting in a similar manner. Apart from the fact that Kobert obviously considers the binding of the ricin to lipoid structures of the cell membrane to be an initial step of the ricin effect, these re-

1   Elfstrand literally speaks of blood corpuscle agglutinating proteins (Elfstrand 1898) ("Blutkörperchen – agglutinierende Eiweisse").

sults are in agreement with modern findings. In some other cases Kobert maintains his old ideas against critical opinions.

– Homogeneity of ricin

A number of authors (Cushny 1898; Liebermann 1907; Müller 1898) doubted that Stillmark's and Kobert's ricin is a uniform substance. This has certainly been declared most clearly by L. von Liebermann (1907). "In my opinion it is the simplest thing to assume that in the rizin there are an agglutinin and a toxin of general effect side by side." Kobert's answer to this is: "The assumption that rizin is a mixture of agglutinin and toxin and that the blood corpuscles mainly anchor only the agglutinin but not the toxin … is incorrect." Kobert, however, was not the only one holding this opinion. Oppenheimer (1904), for example, believes that the ricin on sale contains three reactive groups: a toxophorous, an agglutinophorous, and a haptophorous group.

– Binding site of ricin

Up to that time, neither Kobert nor any other authors had stated an unambiguous formulation suggesting a binding of the plant hemagglutinins to carbohydrate structures of the cell membrane. Especially lipoids are discussed as a site of action. Kobert believes that there are two or three phases of the ricin effect: In the first phase ricin becomes attached to the cells whereas in the second phase the lipoid membrane is transformed into "ricin lipoid", which "physically has a pronounced tendency to get tightly attached to other structures of the same kind in an inseparable way and to fuse with those, losing their own form." In a third phase, there may be, according to Kobert, an occurrence of a hemolysis in case of a large excess of ricin.

L. von Liebermann (1905) reduces the ricin effect to an acid/base reaction. In this, ricin represents the acid whereas the stroma of the erythrocytes functions as the base.

Raubitschek (1909) found that it is possible to deagglutinate blood corpuscles that had been exposed over many hours to the action of a plant agglutinin (he used ricin, abrin and extracts of beans, lentils, peas, sweet peas and many kinds of *Datura*) and are completely agglomerated, by adding a peptone solution or a normal serum, and in this way he obtained a fully homogenous suspension of blood corpuscles. Nobody had the idea yet that this inhibitory effect was caused by the glycoconjugate nature of the serum proteins or by the sugar content of the peptones. It is by the way very interesting that Raubitschek's article is provided with the subtitle *On Healing Attempts in the Test Tube*. The conception that the toxic effect of ricin and abrin was based on a primary action on the erythrocytes was obviously still widespread at the time. After all Raubitschek restricted it by saying: "At any rate it is a matter of opinion whether the above described experiments can rightly be called healing attempts since agglutination does not mean a destruction of the erythrocytes as it is the case in hemolysis."

– The mechanism of the toxic effect

Kobert, too, arrives at the result that 50 to 100 times diluted blood is agglutinated by ricin, whereas "undiluted blood is protected by its plasma from the effect of the poison". "In this way the poison gets with the blood into all organs and is obviously anchored in the important cells of the central nervous system which are protected by an abundant amount of surrounding plasma and are paralysed by it. It does not necessarily produce major anatomic changes here." The changes of the small intenstine which are especially observed in herbivores are, according to him, due to agglutination, which he however could not prove reliably microscopically. He moreover assumes that the poison is excreted through the mucous membrane of the intestine. Oppenheimer (1904) also emphasizes that there is no occurence of hemagglutination in vivo.

– Enzymatic nature of ricin

The classification of ricin as an enzyme by Stillmark unleashed an intensive controversy of opinions, which was not finished yet in 1913. In his work *Are Toxins Enzymes?* L. von Liebermann (1905) arrives at the conclusion that the agglutinating effect of ricin is on no account enzymatically conditioned (which is in full agreement with our present conceptions).

Oppenheimer (1904) on the other hand considers ricin, which by the way in his opinion is not of a protein nature, to be a compound showing many analogies to enzymes. In connection with L. von Liebermann's study, Kobert draws the attention to a ricinus lipase that, though not identical with ricin, can exert an agglutinating effect similar to ricin. Closer relations between the effects of lipase and ricin were not excluded by Kobert.

## 1.6 Concluding Remarks

We have seen how experiments carried out by Bower, Werner, Ritthausen and Dixson in the final analysis culminated in the preparation of ricin by Stillmark. His most important achievement undoubtedly consisted in practising the method designated by him as "pharmacological mode of preparation." First he considered the toxicity, later, after he had discovered this extraordinarily important effect, also the erythrocyte agglutination to be the criterion of his successful experiments of the isolation and purification of ricin. The recognition of the hemagglutination capacity of *Ricinus communis* extracts represented the first step toward the creation of lectinology. Moreover, the study of the so-called phytohemagglutinins surely made easier the understanding of hemagglutination by antibodies (isohemagglutinins). The experienced pharmacologist Kobert from the very beginning assessed the importance of this substance correctly. His activity and the cooperation with the firm of Merck in Darmstadt initiated by him paved the way for Paul Ehrlich's experiments, which created the foundations of immunology. So there exist between immunology and lectin research besides objective connections (possible relationship under consideration of evolutionary connections, similar chemical modificability), also historic relations. One may be surprised how long it took until the real lectin nature, the capability of specific sugar binding, was recognized. Instead, the tendency was prevailing over a long time to see the primary reaction of the lectins in a binding to lipoid structures.

It is also interesting to see to what extent the early studies have exerted an influence up to our present time. This is at least true of the definition of the term of lectin. The postulate of a number of authors that the capacity of cell agglutination is one of the cardinal properties of lectin can be doubtlessly traced back to Stillmark and Kobert. In the last 10 to 15 years ricin research has brought new and surprising results. Above all the investigations of Olsnes (Olsnes and Pihl 1978) have helped to clarify the mechanism of the ricin toxicity. According to these findings, the ricin effect is based on the fact that after the binding of the B-chain to D-galactose structures on cell membranes, the A-chain (see Scheme 1) gets into the cell via a specific mechanism and can inhibit there the protein synthesis on a ribosomal level. Besides ricin and abrin also other toxic lectins, which were discovered later, are capable of producing this effect (modeccin, mistletoe lectin). In a number of plants A-chains are also present in a free form. When we have to state that Stillmark and Kobert had to leave a number of questions unanswered (e.g. the composition of their ricin, binding mechanism of this substance), we must not forget that ricin even today presents a number of problems to us. For example, we do not know what biological function ricin and the other toxic lectins have and which is the substrate of the enzymatically active A-chain. Nor has up to now a decision been made as to what therapeutical importance the so-called immunotoxins (Vidal et al. 1985) will have in future. One can imagine that these synthetic products are derived from ricin by a replacement of the little specific B-chain by (monoclonal) antibodies against selected cell markers.

For more than a hundred years Stillmark's and Kobert's research work has had a directive effect and placed whole generations of scientists before problems part of which could be solved by them and others that have remained unsolved until today. In this connection two aspects of Stillmark's work should be pointed out, which obviously can be fully appraised only today:

1) He explicitly strove to treat his experimental animals with great care. He succeeded in doing so by using the hemagglutinating ricin effect for testing the activity instead of the animal toxicity. One can regard it as the luck of a clever man that he obtained additional findings in this way.
2) Stillmark's remarks regarding the enzyme nature of ricin quoted in this article have proved to be an ingenious hypothesis which a hundred years later is no doubt basically accepted with respect to the protein synthesis inhibiting effect of the A-chain but which still needs final confirmation.

Certain stages of the development of a field of sciences reflect at once all its problems. The "archaic" phase of the ricin research dominated by Stillmark's and Kobert's work directed interest to basic questions of lectin research and was paving the way for the development of lectinology.

# 1.7 References

Bower H (1854) On ricinus communis. Am J Pharm 36:207–209

Cushny AR (1898) Über das Ricinusgift. Arch exp Path Pharm 41:439–448

Dixson T (1887) Ricinus communis. Australasian Med Gaz 8:137–138, 155–157

Ehrlich P (1891a) Experimentelle Untersuchungen über Immunität I Ueber Ricin. DMW 17:976–979

Ehrlich P (1891b) Experimentelle Untersuchungen über Immunität II Ueber Abrin. DMW 17:1218–1219

Elfstrand M (1898) Über blutkörperchenagglutinierende Eiweisse. In: Kobert R (ed) Görbersdorfer Veröffentlichungen a. Band I, Enke, Stuttgart, pp 1–159

Funatsu G, Kimura M, Funatsu M (1979) Primary structure of Ala-chain of ricin D. Agric Biol Chem 43:2221–2224

Hellin H (1891) Der giftige Eiweisskörper Abrin und seine Wirkung auf das Blut. Thesis, Dorpat

Kobert R (1913) Beiträge zur Kenntnis der vegetabilischen Hämagglutinine. Die landwirtschaftlichen Versuchs-Stationen 79:97–151

Lau C (1901) Über vegetabilische Blutagglutinine. Thesis, Dorpat

Liebermann Lv (1905) Sind Toxine Fermente? DMW 31:1301–1305

Liebermann Lv (1907) Über Hämagglutination und Hämatolyse. Arch Hyg 62:277–342

Lugnier AAJ (1974) Isolement et étude des propriétés physicochimiques, biochimiques et biologiques de la ricine, protéine toxique du ricin (ricinus communis L). Thesis, Strasbourg

Müller F (1898) Zur Toxikologie des Ricins. Arch exp Path Pharm 42:302–322

Olsnes S, Pihl A (1978) Abrin and ricin – two toxic lectins. TIBS 3:7–10

Olsnes S, Pihl A (1982) Toxic lectins and related proteins. In: Cohen, van Heyningen (eds) Molecular Action of Toxins and Viruses. Chapter 3. Elsevier Biomedical Press, Amsterdam, pp 51–105

Oppenheimer C (1904) Toxine und Antitoxine. Fischer, Jena

Raubitschek H (1909) Zur Kenntnis der Hämagglutination Über Heilversuche im Reagenzglas. Wiener Klin Wochenschr 22:1065–1066

Reid G (1913) Beiträge zur Kenntnis der chemischen Natur und des biologischen Verhaltens des Rizins. Die landwirtschaftlichen Versuchs-Stationen 82:393–414

Ritthausen H (1882) Zusammensetzung der Eiweisskörper der Hanfsamen und des kristallisirten Eiweisses aus Hanf- und Ricinussamen. J f prakt Chemie 25:130–137

Rosbach HJ (1969) Die Entwicklung des Pharmakologischen Instituts Rostock von seinen Anfängen bis zur Gegenwart und die Bedeutung für die medizinische Fakultät, Teil I. Thesis, Rostock

Stillmark H (1888) Ueber Ricin, ein giftiges Ferment aus den Samen von Ricinus comm. L und einigen anderen Euphorbiaceen. Thesis, Dorpat

Stillmark H (1889) Über Ricin. In: Kobert R (ed) Arbeiten des Pharmakologischen Instituts zu Dorpat, Enke, Stuttgart, pp 59–151

Vidal H, Casellas P, Gros P, Jansen FK (1985) Studies on components of immunotoxins: purification of ricin and its subunits and influence of unreacted antibodies. Int J Cancer 36:705–711

Werner E (1870) Ueber "Ricinin" und den wirksamen Bestandteil der Ricinussamen. Pharmaceutische Zeitschrift für Russland 9:33–44

## Additions in Print

In the meantime Endo and his colleagues have shown that the A-chain of ricin has a RNA N-glycosidase specificity (Endo and Tsurugi 1987; see also Olsnes 1987).

The A-chains of abrin and modeccin have the same specificity. Also the A-chain of mistletoe lectin I from *Viscum album* acts in the same manner (Endo et al. 1988). Yaeta Endo will give a review about his investigations in the next volume of "Advances in Lectin Research".

## References

Endo Y, Tsurugi K (1987) RNA N-glycosidase activity of ricin A-chain. Mechanism of action of the toxic lectin ricin on eukaryotic ribosomes. J Biol Chem 262:8128–8130

Endo Y, Tsurugi K, Franz H (1988) The site of action of mistletoe lectin A-chain on eukaryotic ribosomes: The RNA N-glycosidase activity of the protein. (in press)

Olsnes S (1987) Closing in on ricin action. Nature 328:474–475

# 2 Preparation of Plant Lectins
Harold Rüdiger

## 2.1 Introduction

The existence of lectins was detected already one century ago through their ability to agglutinate erythrocytes. For the early history see Tobiška (1964). Lectin purification generally followed the development in protein chemistry. It began with precipitation methods, and later on chromatographic procedures based on ion exchange and gel filtration were applied to lectins. A great advance came when in the 1960's Agrawal and Goldstein (1967) introduced affinity chromatography to the purification of lectins.

From then on, most lectins have been purified by the use of affinity chromatography, and only in cases where the binding carbohydrate is unknown or not available conventional procedures are still in use.

In 1978, Goldstein and Hayes (1978) published an extensive review on individual lectins. Three years later, Lis and Sharon (1981) compiled affinity chromatography methods applied to the purification of lectins. The present review will therefore primarily deal with more recent papers. In a first chapter, general methods for preparation of plant material and for lectin isolation will be summarized, after which the purification of individual lectins will be described. Some toxic proteins will also be included as far so they are related to know lectins, in spite of the fact that their designation as lectins is disputed (Goldstein et al. 1980; Kocourek and Hořejší 1981). Only those lectins will be listed which have been isolated to homogeneity and characterized with respect to molecular weight and subunit composition while agglutinating activities in more or less crude preparations are not mentioned unless special reasons justify it. Animal, algal and bacterial lectins and mitogens will not be covered. Lack of space also prevents us from giving molecular and other properties of the lectins throughout.

As far as possible, the English, German, and French names of the plants are given.

## 2.2 General Part

In the following, some general rules are outlined how to purify lectins. Only a few references will be given, because experiences of this kind are hardly found in the literature and originate mostly from our work. Since lectins usually are isolated from seeds, mostly from Leguminosae, special emphasis will be given to this starting material.

All recommendations, however, can be transferred to other plant material after proper adaption.

They can be used correspondingly also for the isolation of lectins from bacterial and animal sources.

## 2.2.1   Pretreatment of Plant Material

### 2.2.1.1   Dry Seed Meal

Normally, seeds may be ground in the dry state. This only works well if they are not too hard.
For many seeds a usual coffee-grinder will suffice; for harder seeds a mill with a breaker is
necessary. Grinding of dry seeds is preferable to processing swelled seeds because the proteins
remain in a native state without secondary changes which may occur on swelling. It is, however,
a disadvantage in this method that fragments of the outer seed layers are difficult to remove
from the dry seed meal. These fragments do not contain lectin but often are rich in substances
which disturb the following purification steps. In particular, dyes or substances that are readily
oxidized to dyes may be present in the seed coat. They often deteriorate column materials and
accompany lectin preparations even up to the final purification steps.
Further trouble may arise from soluble polysaccharides (galactomannans) of the seed endosperm which take up water from the extraction buffer. This may lead to highly viscous solutions
which interfere with precipitation and chromatographic steps.
A very effective method to get rid of colored materials is extraction of the dry seed meal with organic solvents (methanol, methylene chloride). This at the same time helps to remove lipids
which are present in large amounts in many seeds.

### 2.2.1.2   Swelling of Seeds

Instead of grinding, the seeds may also be soaked in water or in aqueous buffer for some hours.
Since swelling is the initial step in germination, this in principle may lead to proteolytic degradation of proteins. We have observed such a decay only with storage proteins, but not with lectins which generally are fairly resistant to proteolytic attack. Seeds which do not take up water
if imbibed have either to be ground in the dry state or to be damaged artificially in order to facili-
tate the entrance of water.
If the seeds are not too small, it is recommendable to remove the seed coat prior to homogenization and, if possible, also the endosperm. By this, the complications by colored or viscous substances mentioned above can be circumvented. Losses in lectin do not occur, because leguminous lectins generally are localized in the cotyledons.
The swelled plant material is easily homogenized in a household mixer, an Ultraturrax, a War-
ing blender or similar instruments.

## 2.2.2   Extraction

Extraction is often simply done with water or with sodium chloride solutions (saline). This may
be disadvantageous because extraction of plant material often liberates acidic substances.
Many plant proteins, however, do not dissolve even at only slightly acidic pH. So one runs the
risk that a considerable portion of the protein remains undissolved. This portion may entrap soluble substances and prevent them from going into solution.
In our experience, it is better to use buffers at pH values above 7 for extraction rather than unbuffered saline. If necessary, the pH should be adjusted after homogenization.
As a buffering substance, Tris(hydroxymethyl)aminomethane (Tris) of the lowest quality may
be used. Phosphate or pyrophosphate, though even cheaper, is disadvantageous for the follow-

ing reason: Many lectins need $Ca^{2+}$ and other bivalent ions for full activity. Since these ions bind only weakly to lectins, they are easily removed from the lectin by phosphate or pyrophosphate. Severe activity losses have been observed after extraction with phosphate or pyrophosphate. Though the activity may be restored by supplementing the extract with bivalent cations, extraction with ion scavenging buffers should be avoided.

All buffers have to be preserved against microbial growth. Usually 0.02 % $NaN_3$ is recommended, but in crude extracts which are rich in nutrients for microbia, and in azide-reactive substances, higher concentrations may be necessary.

Depending on the price of the plant material and the time available, the pellet from the first extraction may be reextracted a few times in order to enhance the yield. If the lectin is to be isolated on a laboratory scale, the soluble extract is separated from the debris by centrifugation at moderate speed. For larger batches, filter presses may be used.

## 2.2.3 Prepurification of the Extract Prior to Chromatography

Usually it is unsuitable to apply crude extracts to columns.

First, crude extracts contain vast amounts of nonlectin proteins which will occupy much of the absorbent's capacity. Most of them belong to the storage proteins. These proteins generally are very susceptible to proteolysis, even if no external proteases are added. Their fragments, however, are often less soluble than the intact proteins. Thus, on applying crude extracts to columns, some of the storage proteins will be cloven during chromatography, their fragments will precipitate on the column and finally clog it.

Second, unprocessed crude extracts contain low molecular weight substances. Some of these may be sugars or glycoconjugates which fit to the sugar binding site of the lectin to be purified. Though usually the concentrations of these glycoconjugates are not high enough to prevent the lectin from being adsorbed to the affinity gel, they may reduce its capacity.

As a prefractionation method, many workers use precipitation with ammonium sulfate. After this, the lectin containing precipitate is resolubilized and dialyzed. This reliably removes low molecular weight substances which may interfere with affinity chromatography. Storage proteins, however, are removed only incompletely and will therefore interfere with the following chromatographic steps.

In our experience, it is much more simple to subject the crude extract to a gentle acid precipitation by adjusting it under stirring and pH control with 1 M acetic acid to a final pH of 5.0. At this pH, most lectins remain in the supernatant, whereas most of the storage proteins precipitate. The acidified extract is centrifuged and readjusted to a neutral pH with a diluted (1 M) solution of sodium hydroxide as soon as possible since some lectins do not tolerate low pH values for a longer period of time.

If time has to be saved and high yields are not important, extraction of the seed meal and acid precipitation can be done in one step with only one final centrifugation.

The acid soluble portion of the crude extract is now readjusted to a neutral pH and ready to be applied to a column. It is, however, advantageous first to concentrate the solution in order to remove low molecular weight substances and to save time during application to the column. Freeze drying is not generally recommended because some lectins are not easily resolubilized. Ultrafiltration, either simply by applying a water pump vacuum to a dialysis tube filled with the protein solution or by one of the more sophisticated commercial variants of this technique is preferred because the lectins remain dissolved. Finally, the material should be dialyzed in order

28

to further reduce the concentration of possible inhibitory glycoconjugates, centrifuged at high speed and applied to a column.

## 2.2.4 Chromatographic Purification

All methods suitable for protein purification may also be used for lectins: ion exchange, gel filtration, isoelectric and chromatofocusing.

Since lectins are able to specifically bind carbohydrates and glycoconjugates but in contrast to enzymes do not convert them chemically, the method of choice for lectin purification is affinity chromatography with immobilized carbohydrates (mono-, oligo-, polysaccharides or glycoconjugates).

## 2.2.4.1 Affinity Chromatography by the Use of Polysaccharides

Lectins specific for glucose (and, as far as leguminous lectins are concerned, at the same time also for mannose) bind to glucose polymers that are connected α-glycosidically. This has been described already as early as 1936 by Sumner and Howell (1936) who observed precipitates on mixing concanavalin A (Con A), the lectin from *Canavalia ensiformis* beans with glycogen and starch solutions. The first to utilize this binding ability for the purification of a lectin were Agrawal and Goldstein (1967), and Olson and Liener (1967) who employed the commercial gel filtration medium Sephadex, which is a chemically cross-linked dextran, to purify Con A from crude extracts. This method works well also with other lectins of similar sugar specificities, as, e.g., from many *Vicia* species, from *Lens culinaris* and from *Pisum sativum*, though capacities for individual lectins are different. Differences in capacity are also found between the various Sephadex types, depending on the degree of cross-linkage. Highly cross-linked types do not bind lectins because of their narrow-pores. Wide-pore types generally should be expected to have higher capacities than narrow-pores ones, but due to swelling they contain less glucose units per milliliter and are more difficult to handle than the more rigid narrow-pore types. We found a capacity optimum around the types Sephadex G-75 and G-100.

If large amounts of lectin have to be prepared, and commercial gels are not available in the required quantity, dextran (high-molecular weight type) may be cross-linked with divinylsulfone according to the method of Young and Leon (1978). The resulting gel lump is broken up cautiously, e.g., by pressing it through the holes of a Büchner funnel. Fines are removed by repeated decantation, and the easily sedimenting particles are used either in a short column or in a batch procedure.

Lectins specific for galactose can likewise be purified using agarose (a copolymer of D-galactose and 3.6-anhydro-L-galactose). This has been done with the lectins from *Abrus precatorius*, *Ricinus communis* and *Arachis hypogaea*. Commercial agarose beads (4% and 6% agarose) have been used successfully.

For some galactose binding lectins, the capacity of native agarose is very low or even absent. In these cases it can be tried to improve the capacity by subjecting the agarose to a gentle acid treatment [e.g., 4h, 0.2 M HCl, 50°C, (Ersson, et al. 1973; Allen and Johnson 1976)], which partially hydrolyzes glycosidic bonds of the polymer and exposes more binding sites to the lectin. The lectins from *Bauhinea purpurea*, *Crotalaria juncea*, *Glycine max*, *Sophora japonica* and *Wisteria floribunda* bind only to acid-treated, not to native agarose.

29

Lectins with a specificity for N-acetylglucosamine (GlcNAc) may be purified on chitin which is a GlcNAc polymer. This has been done with lectins from *Griffonia simplicifolia* (Shankar et al. 1976), *Urtica dioica* (Peumans et al. 1984a), *Chelidonium majus* (Peumans et al. 1985a) and some others.

Several lectins, as from *Arachis hypogaea*, *Griffonia simplicifolia* (Young and Leon 1978), *Sophora japonica*, *Wisteria floribunda* (Freier 1985), can also be purified using divinylsulfone-cross-linked plant gums (locust, quar gum) which consist of galactomannans.

In nearly all cases, desorption is accomplished by a hapten sugar. In the case of the mannose/glucose binding lectins, glucose will suffice, though α- glycosides of glucose or mannose are more effective. Lectins binding to agarose are desorbed with galactose or the less expensive lactose.

Generally, lectins can also be desorbed with acidic buffers (ph between 5 and 3). This, however, should only be done if the lectin is acid stable.

Isolectin forms or lectin fragments have been separated from each other by desorption with gradients instead of sudden concentration changes (Rüdiger 1977; Obata et al. 1978; Fleischmann et al. 1985).

## 2.2.4.2 Affinity Chromatography by the Use of Immobilized Sugars

In cases where the binding sugar is not available as a polymer, monomeric, oligomeric or some other carbohydrate derivatives have to be immobilized to a matrix.

Due to the low reactivity of hydroxy groups at moderate pH, there are not many methods available to immobilize monosaccharides. High pH values, on the other hand, may lead to rearrangements of the carbohydrates. Additionally, immobilization of a carbohydrate directly to a matrix carries the risk of sterical hindrance which prevents binding of the lectin.

Direct immobilization of carbohydrates can be achieved by the use of butanediol-bis-glycidylether (bisoxirane method) or with divinylsulfone as coupling agents (Porath 1974). Both procedures need high pH values (pH around 11, 12).

A less drastic coupling procedure can be used if the lectin is specific for an acetylamino sugar. Allen and Neuberger (1975) coupled galactosamine to a carboxyl croup bearing matrix (CH-Sepharose) and found that the GalNAc specific lectins from *Glycine max* bind to such gels, in spite of the fact that the amino group of the binding sugar does not bear an acetyl but a longer chain ($C_6$) acyl residue.

A further method for immobilizing monosaccharides was introduced by Hořejší and Kocourek (1973). These authors prepared adsorbents by copolymerizing allylglycosides with acrylamide. Though this method appears to be generally applicable, wider use has presumably been prevented by the effort in preparative work and by the fact that the gels are not in a bead shape.

Many adsorbents carry sugar residues via a spacer arm. Bloch and Burger (1974) described the preparation of an affinity adsorbent in a simple one-pot reaction. First the nitro group of a nitrophenylglycoside is reduced, then the resulting aminophenyl glycoside is coupled to a carboxyl bearing matrix (CH-Sepharose) by means of a water-soluble carbodiimide. This adsorbent works well with some lectins. The introduction of a phenyl group as a spacer, however, may lead to hydrophobic interaction with the lectin. We found (Rüdiger 1977) that desorption of the *Vicia cracca* lectin from this adsorbent requires much more drastic methods than from the adsorbent of reference (Allen and Neuberger 1975).

Gordon et al. (1972) describe the preparation of a gel which contains the sugar bound via an ε-aminocaproyl spacer arm. A very simple but lengthy method for preparing an adsorbent with a spacer bound sugar is by reductive amination of disaccharides. Baues and Gray (1977) and Matsumoto et al. (1981) started from a matrix containing primary amino groups. It was first reacted with disaccharides (e. g., maltose, lactose), and after a proper time (1, 2 weeks) the Schiff bases formed were reduced with cyanoborohydride. The resulting gels have high capacities.

## 2.2.4.3 Affinity Chromatography by the Use of Immobilized Glycoconjugates

Usually plant extracts are screened for lectins with red blood cells, either of human or of animal origin. Lectins interact with glycoconjugates exposed to the outer cell surface. So in principle erythrocyte membranes or their glycoconjugates after immobilization should form suitable affinity adsorbents. This has been tried by some workers (for details see Lis and Sharon 1981), but the adsorbents appear to suffer from low capacities. Affinity chromatography with pure glycoproteins or glycopeptide fragments from them has been more successful (Scheme 2.1). Many papers deal with the isolation of individual lectins by the use of immobilized animal and human blood group active proteins, immunoglobulins, thyroglobulin, fetuin, and ovomucoid, which offer a host of various carbohydrate structures. We found that the capacity of such adsorbents is greatly improved if prior to immobilization the glycoconjugates are discharged from their terminal sialic acid residues. This can be done by a gentle acid treatment. By using immobilized desialylated glycoproteins, we could isolate lectins from nearly 30 plant species (Freier et al. 1985).

In contrast to matrix-bound carbohydrates, the carbohydrate structures of immobilized glycoproteins are easily accessible to the lectin, because the protein serves as a spacer arm. Further-

---

*Scheme 2.1*   Affinity adsorbents used for the purification of lectins

– Naturally occurring polysaccharides, either in their native state or chemically cross-linked

– Mono- or oligosaccharides, immobilized to a matrix
  • by the bisoxirane method
  • by the divinylsulfone method
  • via an amide group (for amino sugars)
  • by copolymerization of allylglycosides and acrylamide
  • via a spacer group
  p-aminophenylglycosides
  p-aminobenzyl-thioglycosides
  ε-aminocaproyl-N-glycosylamines
  di- or oligosaccharides, immobilized via Schiff base formation and reduction, using the terminal sugar as a spacer

– Immobilized glycoproteins
  • stabilized red blood cells, yeast cells
  • glycopeptides from red blood cell membranes
  • glycopeptides from hog gastric mucin
  • glycoproteins: fetuin, thyroglobulin, immunoglobulins, ovomucoid, blood group substances, submaxillar or gastric mucins.

---

more, in contrast to carbohydrates a broad arsenal of gentle immobilization methods is available for proteins. Mostly, proteins are coupled via reactive primary amino groups by the cyanogen bromide procedure originally introduced by Porath and co-workers (Porath 1974). This method together with others is described and exemplified by detailed instructions in a recent monograph (Dean et al. 1985). A very effective modification of the procedure of Porath not included in reference (Dean et al. 1985) has been published by Kohn and Wilchek (1982). In the Porath procedure, high pH values are necessary to form a reactive species from agarose and cyanogen bromide. This, however, leads to side reactions which consume most of the cyanogen bromide and are difficult to control. In the method of reference (Kohn and Wilchek 1982), activation of the gel is not performed in alkaline solution, but at a moderate pH at low temperature. This greatly reduces the amount of cyanogen bromide necessary and makes the procedure very reproducible.

Mostly the success of immobilization is not measured directly but inferred from the success in isolating a lectin. This is because most methods for measuring the ligand concentration are neither direct nor precise. Ligand concentrations can be calculated from the difference between the amounts employed and recovered. If a radioactive ligand is used, the degree of immobilization can be measured directly, but this is not a routine method. The same is true for methods where the gel is destructed and its constituents brought into solution for measurement. As far as proteins are concerned, advantage may be taken of their UV absorption. Though this has been tried by several workers, it is difficult to overcome the intense UV light scattering of gels, in particular after reaction with cyanogen bromide which increases the opacity. Some years ago we developed a simple method (Schurz and Rüdiger 1982) for measuring the protein density of affinity gels directly. In this, the protein-loaded gel is filled into a 1 mm cuvette and allowed to settle. Then the first derivative of the spectrum is recorded and compared with values from a standard solution of the particular protein. By this, the steep background absorption of gels in the UV range, especially after the reaction with cyanogen bromide, is eliminated. If the background is not sufficiently linear in the relevant range, the second derivative spectrum may be recorded. Since most modern spectrophotometers offer the possibility to record derivative spectra, this is a simple routine method.

In the visible range, the background by light scattering from the gel is much smaller. Therefore after staining the gel-bound protein a normal spectrum may be recorded (Schurz and Rüdiger 1982; Asryants et al. 1985).

In Figure 2.1, a running scheme for the purification of lectins is shown. Scheme 2.1 summarizes the preparation of affinity adsorbents.

## 2.2.5    Characterization of Lectins

## 2.2.5.1  Chemical Characterization

The standard method for characterizing proteins with regard to molecular weight is electrophoresis in polyacrylamide gels. Several review articles and monographs deal with this topic (e.g. Maurer 1971), therefore details will not be given here.

Subunit molecular weights are determined by electrophoresis under dissociating conditions, mostly following the procedure of Laemmli (1970) in which the sample is heated in the presence of urea, mercaptoethanol and sodium dodecyl sulfate, and the electrophoresis runs in the presence of sodium dodecyl sulfate. The total molecular weight of native proteins can also be determined electrophoretically, e.g. by a run in a polyacrylamide concentration gradient

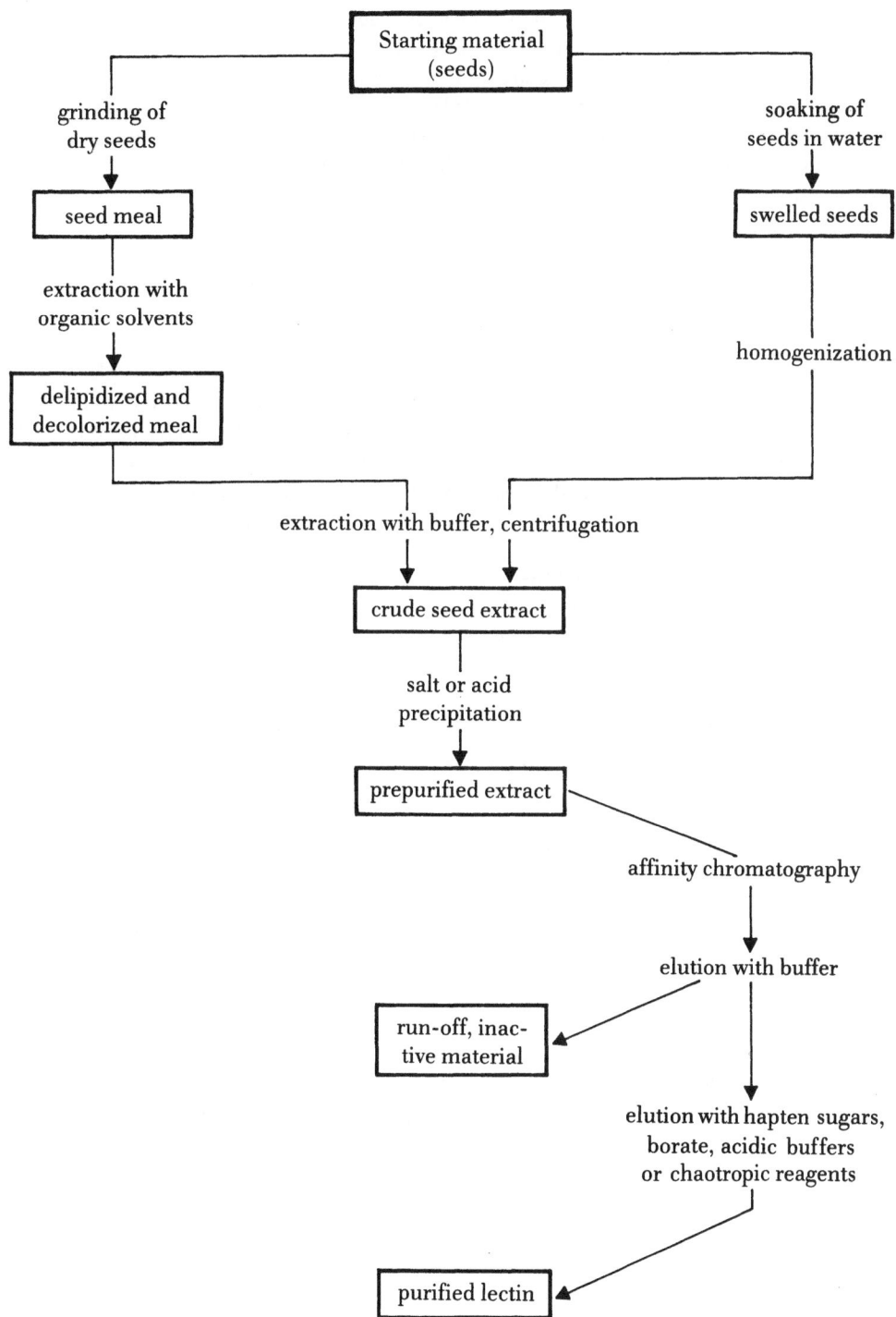

*Fig. 2.1* General scheme for the purification of lectins

(Campbell et al. 1983). Determination of molecular weight by gel filtration has to be looked at with caution, even if the gels used are not suited as affinity adsorbents for the particular lectin. Weak interactions that only lead to retardation may pretend seemingly low molecular weight values (Rüdiger 1977; Young and Jackson 1984). Amino acid sequences of many lectins have been determined. Though sequence determination is not a routine method, it should be pointed out that many lectins are highly homologous (Chapot et al. 1986; Strosberg et al. 1983), in spite of the fact that a common biological role of these proteins is still a matter of conjecture.

## 2.2.5.2  Characterization by Biological Activities

The most common method to trace lectins and to determine their concentration at least semiquantitatively is hemagglutination. For this purpose, usually commercial plastic titer plates which contain 100 µl-wells with a V-shaped bottom are used. Within a row of wells all except the first one are filled with 25 µl of an isotonic buffer (e. g. 0.05 M Tris/HCl pH 8.0; 0.1 M in NaCl; Tris is preferred to phosphate for the reason mentioned above), then 25 µl of the lectin solution to be analyzed is filled into the first two wells, and from the second well a serial dilution with 25 µl is started. Finally, each well is provided with 25 µl of a freshly prepared 2%, 4% red blood cell suspension in the isotonic buffer and the plate incubated for at least 30 min, after which the titer is read visually. Agglutination manifests itself by an evenly distributed layer of red cells over the whole bottom, in its absence the red cells collect only at the apex of the well. The titer of the original solution is equal to the dilution in the last well to show agglutination (number "n"), i.e. titer = $2^{(n-1)}$. E. g., if up to the fourth well the cells are agglutinated, the titer is $2^{(4-1)} = 8$.

It is recommended not to dilute in too many steps on a plate. Dilution errors are inevitable and progress in a geometrical order. We usually prefer to predilute on a larger scale (about 1 ml) by factors of 10, 100, and 1000. Then these solutions are further diluted in triplicates on a plate by no more than 4 to 5 steps.

If these precautions are followed, titer determinations with the same blood sample are quite reproducible and much better than only ±1 dilution step as usually claimed.

Blood cells from humans and animals can be used. Outdated human blood conserves are available from transfusion centers. Some lectins show an absolute specificity in the AB0(H) system, others only small (2, 3 dilution steps) or no differences between the blood groups. Therefore, blood of the groups A, B and 0(H) should be kept in stock. Blood conserves kept in the refrigerator are suited over months. Finally, however, hemolysis occurs and the titers tend to rise, presumably because of slow hydrolytic demasking of surface structures. In the diluted suspension used for titer determination, however, cell breakdown is very fast. Therefore, fresh suspension has to be prepared at least every day. Blood cells are more stable toward hemolysis after treatment with glutaraldehyde (Turner and Liener 1975).

Not all lectins respond to native red cells. Some require subterminal structures to be exposed. The lectin from *Arachis hypogaea* reacts only with cells treated with neuraminidase. For many lectins, the titers are higher with trypsin or papain-treated cells. Some animal cells, as e.g., from rabbits, often display higher titers than human cells. High titers can also be achieved in many cases if dilution is performed in a buffer which contains proteins (serum albumin) or polyvinylpyrrolidon solutions. Occasionally, the addition of $Ca^{2+}$ ions may also enhance the titers. Since the agglutination determination in the simple form described is not very precise, modifications have been proposed in which not the agglutination itself but rather optical density of the nonagglutinated red cells is measured (Turner and Liener 1975; Lis and Sharon 1972).

Also precipitation of glycoconjugates by lectins as measured by turbidimetry, nephelometry or some other parameter have been used to follow lectin concentration and activity (for details see Goldstein and Hayes 1978).

Many lectins are able to stimulate mitosis and proliferation of animal and human lymphocytes. The technique to determine lymphocyte proliferation is more complicated than the agglutination assay with which, however, it shares the inaccuracy and the deviations found between different cell charges. This absolutely necessitates to carry along standard mitogens with each series of measurements. Only a brief description of the method will be given here, for details see reference (Gebauer et al. 1982) and the papers cited there.

Mouse cells are taken from the spleen, from humans peripheral lymphocytes are used. The cells are diluted to a density of about $2 \cdot 10^6$ cells per ml. A concentration series of the mitogenic lectin is pipetted into the U-shaped wells of a plastic microtiter plate. Each well is provided with 3 to $4 \cdot 10^4$ cells in a medium containing fetal calf serum and antibiotics. The cell cultures are then allowed to grow for 48 h in a moist (95 % relative humidity) atmosphere containing 10 % $CO_2$. After this, the cells are pulsed with radioactive thymidine. This guarantees that only cell proliferation is measured. The cells are incubated for further 18 h, finally harvested by filtration, dried and the radioactivity counted in a scintillation counter. Each measurement has to be performed at least in triplicates. At low lectin concentrations, thymidine incorporation is positively correlated with the lectin concentration, at high concentrations inhibitory effects dominate. Therefore lectins typically display bell-shaped stimulation (thymidine incorporation-vs-log lectin concentration) curves with optimal stimulation in the medium concentration range (2 to 10 µg/ml).

## 2.3  Individual Lectins

Plants marked with * are shown in chapter 4.

**Abrus precatorius** (Leguminosae)*
Jequirity bean, Indian licorice; Paternostererbse
The seeds contain hemagglutinating and toxic activities which belong to two different but related proteins. Special attention has been paid to the toxin as a prospective reagent in tumor therapy.
Olsnes (1978b) and Olsnes and Pihl (1982) reviewed isolation and properties of some toxic lectins, among others those from A. precatorius. Both proteins, the agglutinin (lectin) and the toxin, often referred to as abrin, can be separated from each other by ion exchange chromatography on DEAE-cellulose of the crude extract. Subsequently, contaminating inert proteins are removed from the separated fractions by affinity chromatography on agarose (Sepharose) and elution with galactose. Depending on the seed and on the Sepharose batches, the toxin is not bound but only retarded.
Lin et al. (1981) found four isotoxins two of which bind only weakly to Sepharose and are eluted without galactose, further two toxins and the agglutinin require galactose for elution.

**Adenia digitata** (Passifloraceae , formerly Modecca digitata)
In its roots, this plant contains agglutinating and toxic proteins which are closely related to the toxins from Abrus precatorius and Ricinus communis. Particular interest has arisen for the toxin for the same reason as for abrin. The purification of the toxin has been described by Olsnes et al. (1978; 1982a). They used gel filtration, ion exchange chromatography and finally immobilized

desialofetuin as an affinity adsorbent from which the toxin was eluted with lactose. After these three steps the product is still composed of several isotoxins which differ in pI values.

In the laboratory of Stirpe et al. (1980), the toxin named modeccin was purified by chromatography on agarose. In contrast to other proteins which either run through or are completely adsorbed, the toxin is only retarded. The next step consisted in affinity chromatography on acid-treated agarose and elution by galactose. Retardation of the toxin on agarose has never been observed by Olsnes' group.

Recently, a similar toxin which acts also as an agglutinin, has been isolated from *Adenia volkensii* (Barbier et al. 1984).

## Aegopodium podagraria (Apiaceae)
Ground elder; Geißfuß, Giersch; aegopode

The lectin occurs in the rhizomes. It has been purified by Peumans' group (Peumans et al. 1985b). The only effective mean was a combination of conventional methods with affinity chromatography on cross-linked red blood cells. The protein is extremely large (48,000 MW) and consists of two kinds of nearly equal subunits. It acts unspecifically and is preferentially inhibited by GalNAc.

## Agaricus (Agaricaceae)
edible mushroom; Champignon; champignon

Eifler and Ziska (1980) purified agglutinating proteins by conventional methods. They found two lectins of different molecular weights (60,000 and 32,000 MW, subunits 14,000 MW) which are related immunologically and could not be inhibited by monosaccharides.

Ahmad et al. (1984) also purified the lectins by conventional procedures. Their preparations, however, were nearly equal in molecular weight.

Osawa's group (1985) employed, besides conventional methods, affinity chromatography on immobilized bovine submaxillary mucin. By final isoelectric focusing, they resolved the lectin into four fractions with pI values from 5.5 to 6.7. All lectins displayed the same subunit size (16,000 MW) and were tetramers. As in the work of Eifler and Ziska (1980), no monosaccharide really inhibited, though agglutination was weaker at high α-methyl-GalNAc concentrations.

## Agropyrum repens (Gramineae)
Couch grass; Quecke; chiendent rampant

By Cammue et al. (1985b), a lectin from the leaves has been isolated. It differs greatly from other well-known *Gramineae* lectins which usually occur in the seed embryo (see *Triticum*).

The lectin was isolated on a commercial GlcNAc containing adsorbent. In spite of the isolation on this matrix, it is optimally inhibited by GalNAc.

## Aleuria aurantia (Pezizaceae)
Orange peel fungus; Orangebecherling, pézize

Kochibe and Furukawa (1980; 1982) isolated an L-fucose-binding lectin from the fruiting bodies. In an essential step, the preparation was adsorbed to immobilized porcine submaxillary mucin and eluted with L-fucose. The protein of 72,000 MW is composed of two apparently identical subunits.

## Amphicarpea bracteata (Leguminosae)

Hog-peanut

The seed lectin displays specificity against the human blood group A. It has been isolated by Blacik et al. (1972) on GalNAc, immobilized to agarose by the bisoxirane method.

No molecular data are given.

## Arachis hypogaea (Leguminosae)

Peanut; Erdnuß; arachide

The peanut agglutinin, mostly designated as PNA, does not react with red blood cells unless they have been desialyzed. It binds to galactose residues. In 1975, two methods were described that are still in use even in recent work. As a crucial step Lotan et al. (1975; 1978) used affinity chromatography on immobilized ε-aminocaproyl-β-galacto-pyranosylamine, elution was effected with galactose.

Terao et al. (1975), on the other hand, made use of the weak affinity of the lectin to agarose. After the bulk of contaminating proteins had passed the column, the lectin appeared as a retarded peak without the use of a special desorbent.

Two groups (Baues and Gray 1977; Matsumoto et al. 1981) described the preparation of affinity adsorbents by the reductive coupling of disaccharides to amino group containing gels (reductive amination). Preparation of the gels takes a long time. By immobilizing lactose, a high capacity adsorbent for PNA was obtained.

Sutoh et al. (1977) used guar gum entrapped in polyacrylamide and allyl-galactoside copolymerized with acrylamide for the isolation of PNA. The guar gel had a high capacity.

Cross-linked guar gum and arabinogalactan have been used by two groups (Young and Leon 1978; Majumdar and Surolia 1979).

Uy and Wold (1977) coupled poly- and oligosaccharides to agarose using the bisoxirane procedure. PNA was bound very effectively by the adsorbent prepared from lactose.

Franz and Ziska (1981) used immobilized human immunoglobulins for the isolation of PNA. Newman (1977) found that PNA preparations from various sources are heterogeneous if studied in the native state by electrophoresis or electrofocusing, though they display only one band in electrophoresis under denaturing conditions. Preparative separation of six major and three minor isolectins was achieved by Miller (1983) who used chromatofocusing. Affinity adsorbents prepared by immobilizing desialyzed hog gastric mucin to agarose or to cross-linked dextrans have a broad specificity (Freier et al. 1985). Though their capacity for PNA is only similar to that of Lotan et al. (1975; 1978) and does not match that of adsorbents prepared by reductive amination (Baues and Gray 1977; Matsumoto et al. 1981), it is quickly and easily prepared and can be applied to many lectins.

## Artocarpus (Moraceae)

Jack fruit; Jackfruchtbaum; fruit du jacquier

Interest arose in the seed lectin from A. integrifolia because of its ability to specifically interact with class A immunoglobulins (IgA) (Roque-Barreira and Campos-Neto 1985). The lectin has been purified by Moreira et al. by conventional methods (Moreira and Ainouz 1981), later, Kumar et al. (1982) utilized the specific interaction of the lectin with immobilized guar gum. Affinity chromatography of the seed lectins of two other species on immobilized human blood cell stroma was performed by Moreira et al. (Moreira and De Oliveira 1983) and on a commercial galactosamine-gel by Namjuntra et al. (1985) who also report that their preparation (from A. heterophyllus) cross-reacts immunologically with an antibody directed against the lectin from

*Maclura pomifera* (Moraceae). The preparations obtained from different species and in different laboratories vary in molecular weight and subunit compositions.

### Bandeiraea simplicifolia (Leguminosae)
see *Griffonia simplicifolia*

### Bauhinia purpurea (Leguminosae)
The lectin from the seeds displays a specificity for GalNAc and Gal. It has been purified by adsorption to acid-treated agarose (Allen and Johnson 1976), to immobilized desialyzed bovine submaxillary mucin (Osawa et al. 1978), hog gastric mucin or ovomucoid (Freier et al. 1985), and to a commercial GalNAc containing gel (Young et al. 1985).

### Brachypodium sylvaticum (Gramineae)
False brome grass; Zwenke
Peumans et al (1982a) isolated a lectin from the seed embryos using a commercial gel with immobilized GlcNAc. In several respects, it resembles other *Gramineae* lectins (see *Triticum*).

### Bryonia dioica (Cucurbitaceae)
Bryony; Zaunrübe
An agglutinating activity is present in all parts of the plant except the seeds. From the root stocks, Peumans et al. (1984b) isolated a lectin. Though agglutination can be inhibited by GalNAc and other related simple sugars, attempts to isolate the lectin by the use of immobilized monosaccharides or cross-linked polysaccharides were unsuccessful. The lectin was specifically adsorbed to immobilized fetuin and eluted with lactose, one of the inhibitors of agglutination.

### Butea (Leguminosae)
A seed lectin was isolated by Hořejší et al. (1980) on an acrylamide/allylgalactoside copolymer and by Ghosh et al. (1981a) who used agarose (Bacto-agar) as an affinity adsorbent.

### Cactaceae
Ochoa's group described purification and some biological properties of lectins from several *Cactaceae* (Zenteno and Ochoa 1985). Purification was achieved by a combination of ion exchange, hydrophobic and affinity chromatography on immobilized red blood cell stroma.

### Canavalia ensiformis (Leguminosae)*
Jack bean; Schwertbohne
The lectin from this plant, concanavalin A (Con A), has been known since long (see Introduction). The method of choice for its isolation is affinity chromatography on the easily available cross-linked dextrans (Agrawal and Goldstein 1967; Olson and Liener 1967; Young and Leon 1978). Tailor-made adsorbents, in part with higher capacities, have been proposed (Baues and Gray 1977; Matsumoto et al. 1981; Uy and Wold 1977; Horisberger 1977; Filka et al. 1978). Native Con A consists of the "complete" subunit of 27,000 MW and smaller polypeptides [for the formation of these "fragments" see Carrington et al. (1985)]. Removal of the fragments can be achieved by a simple precipitation procedure (Cunningham et al. 1972) or by affinity chromatography on Sephadex with a glucose gradient for desorption (Cunningham et al. 1972; Obata et al. 1978).

Lectins similar to Con A have also been isolated from related species, as *C. brasiliensis* (Moreira and Cavada 1984) and *Dioclea grandiflora* (Moreira et al. 1983; Richardson et al. 1984).

### Caragana arborescens (Leguminosae)
Pea tree; Erbsenstrauch; arbre aux pois
The isolation of this lectin has already been reviewed in reference (Goldstein and Hayes 1978). In the meantime, it has also been isolated by the use of immobilized desialyzed hog gastric mucin and ovomucoid (Freier et al. 1985).

### Chelidonium majus (Papaveraceae)
Greater celandine; Schöllkraut; chélidoine
Peumans et al. (1985) isolated a GlcNAc-binding lectin from the seeds. As an affinity adsorbent, chitin was used. The lectin appears to be very different from leguminous lectins but to resemble in some respects the lectins from Gramineae, Solanaceae and Phytolaccaceae.

### Cicer arietinum (Leguminosae)
chick-pea; Kichererbse; pois chiche
The seeds have usually been regarded as being devoid of agglutinating activity. Nevertheless, Kolberg et al. (1983) could isolate a lectin which only responds to papainized blood cells. Of the affinity adsorbents tested, the only effective one with a low capacity was immobilized asialo-IgM. The purification procedure proposed by these authors is based on conventional ion exchange and gel filtration steps. No simple sugar, but some glycoproteins are able to inhibit agglutination. The *Cicer* lectin is different from the typical Vicieae lectins to which subtribe *Cicer* has been thought to belong.

### Clitocybe nebularis (Tricholomataceae)
Nebelkappe
By the use of an allyl-α-GalNAc/acrylamide copolymer and Gal as an eluant, Hořejší and Kocourek (1978b) isolated a lectin from the fruiting bodies of this fungus. The protein consists of 19,000 and 14,500 MW subunits which form a 70,000 MW native molecule.

### Crotalaria juncea (Leguminosae)
Sunn hemp; bengalischer Hanf
This lectin, already reviewed in reference (Goldstein and Hayes 1978), has been more recently also purified by the use of cross-linked arabinogalactan (Majumdar and Surolia 1979).

### Croton tiglium (Euphorbiaceae)
Purging croton; Crotonöl-Baum; croton
Two groups describe the isolation of seed lectins (Banerjee and Sen 1981; Fu and Chen 1982) by ion exchange and gel filtration chromatography. In their chemical and biological properties, both preparations appear to be different.

### Cucurbita (Cucurbitaceae)
Pumpkin; Kürbis; courge
From the phloem sap of some *Cucurbita* species, lectins have been found and isolated either by ion exchange and gel filtration (Sabnis and Hart 1978) or by the use of immobilized β-linked oligomeric GlcNAc (Allen 1979; Read and Northcote 1983).

**Cytisus** (Leguminosae)

Broom; Besenginster; genêt à balai

The lectin from *C. sessiliflorus* seeds has already been reviewed in reference (Goldstein and Hayes 1978). From *C. multiflorus*, Osawa's group (Konami et al. 1983a) isolated two lectins. One resembles the *C. sessiliflorus* lectin in agglutinating blood cells of group 0(H) and in binding β-linked GlcNAc oligomers, the other one is not blood group specific and binds to Gal and GalNAc. From *C. scoparius* (formerly *Sarothamnus scoparius*) seeds, Young et al. (1984) isolated two major lectin fractions by adsorption of the extract to a GalNAc gel and by sequential elution with Gal and GalNAc. The fraction desorbed with Gal is similar to the *C. sessiliflorus* lectin in protein structure but not in sugar specificity. The GalNAc-desorbed fraction differs from both. The lectins can also be purified by the use of immobilized desialomucin (Freier et al. 1985).

**Datura stramonium** (Solanaceae)*

Thorn-apple, jimson weed; Stechapfel; pomme épineuse

The lectin from the seeds has been purified by various affinity chromatographic procedures; on a chitin containing polysaccharide mixture from *Aspergillus* (Hořejší and Kocourek 1978a), on fetuin-Sepharose (Kilpatrick and Yeoman 1978), ovomucoid-Sepharose (Desai et al. 1981), and by the use of spacer-bound diacetylchitobiose (Crowley and Goldstein 1981; 1982). Desorption was accomplished with oligosaccharide mixtures prepared from chitin or by acids. The lectin resembles other Solanaceae lectins in carbohydrate specificity, in its content of half-cysteine and hydroxyproline and in the size of its carbohydrate moiety which is rich in L-arabinose.

**Dioclea grandiflora** (Leguminosae)

see *Canavalia ensiformis*

**Dolichos biflorus** (Leguminosae)

Horse gram

From its structure, this lectin is a typical leguminous lectin. It displays a very narrow specificity against GalNAc and its α-glycosides and hence against the human blood group A.

Affinity purification on hog mucin is reviewed in reference (Goldstein and Hayes 1978). Unlike other GalNAc-binding lectins, it is not even retained by adsorbents which contain Gal or N-acyl-galactosamine residues instead of GalNAc (Freier 1985; Lönngren et al. 1976).

The lectin has also been purified by the use of chemically synthesized adsorbents which contain glycosidically bound GalNAc (Filka et al. 1978; Kocourek et al. 1977), on immobilized human immunoglobulins (Franz and Ziska 1981) or desialylated hog gastric mucin (Freier et al. 1985). In addition to the "classic" seed lectin which has been studied essentially by M. E. Etzler's group, further lectins or lectin-like proteins from *D. biflorus* have been detected and isolated by the same workers. A protein cross-reactive with the lectin (called cross-reactive material, CRM) occurs in stem and leaves. In its structure and specificity, it is similar to but not identical with the seed lectin. Though it binds to the same sugar as the seed lectin, it is not a hemagglutinin because interaction with carbohydrates only takes place at low, i.e., hypotonic ionic strengths (Etzler 1983). A third lectin-like protein with a similar carbohydrate specificity but different structure and composition has been detected in roots, stems and leaves of *D. biflorus* plants (Quinn and Etzler 1984).

In a plant from the same genus, *D. lablab*, a seed lectin occurs which is unspecific in the human AB0(H) system. It is inhibited by GlcNAc and α-methyl-D-mannoside and was isolated on immobilized α-linked mannose or ovomucoid as affinity adsorbents (Güran et al. 1983).

40

**Echinocystis lobata** (Cucurbitaceae)

Wild cucumber; Igelgurke

From the seeds, a galactose binding lectin was isolated on cross-linked guaran (Lönngren et al. 1976) and arabinogalactan (Majumdar and Surolia 1979).

**Eranthis hyemalis** (Ranunculaceae)

Winter-aconite; Winterling; éranthe

The root tubers contain a lectin which exhibits a considerable preference for human blood group 0(H) compared with A and B, the best inhibitory monosaccharides being GalNAc and lactose. It was isolated by Peumans' group (Cammue et al. 1985a) who used immobilized fetuin as an affinity adsorbent. Elution was unsuccessful with lactose even at high concentration but could be achieved with water. This suggests that hydrophobic interactions occur between lectin and adsorbent. After two further conventional steps the lectin was pure. It is the first lectin from a Ranunculacea, in its properties very different from the well-known leguminous lectins.

**Erythrina** (Leguminosae)

Coral tree; Korallenstrauch; érythrine corail

From different *Erythrina* species, Gal-binding lectins have been isolated and characterized. Acid-treated agarose served as an affinity adsorbent for the isolation of the lectins from *E. corallodendron* (Gilboa-Garber and Mizrahi 1981) and *E. variegata orientalis* (Datta and Basu 1981). The *E. cristagalli* (Iglesias et al. 1982), *E. edulis* (Perez 1984) and *E. variegata orientalis* lectins (Fukuda et al. 1984) were purified on matrix-coupled Gal, and the *E. indica* lectin on an acrylamide/allyl-α-D-Gal copolymer (Hořejší et al. 1980). The lectins from several *E.* species were isolated by the use of lactose coupled to Sepharose (Lis et al. 1985; Kortt 1986). Though very similar in their chemical properties, the lectins display remarkable differences in their behavior toward human and animal cells.

**Euonymus europaeus** (Celastraceae)*

Spindel tree; Spindelbaum, Pfaffenhütchen; fusain

The seed lectin belongs to those which are not inhibited by simple mono- or oligosaccharides. Its isolation and properties have already been reviewed in reference (Goldstein and Hayes 1978). More recently, it has also been purified on immobilized human immunoglobulins (Franz and Ziska 1981), on desialyzed MN blood group glycoprotein (Petryniak et al. 1981), and on immobilized desialyzed hog gastric mucin (Freier et al. 1985). Desorption from the latter material was effected with GalNAc or borate. From the seeds of *E. sieboldiana*, blood group B-specific lectins were isolated by conventional methods (Yamamoto and Sakai 1981).

**Euphorbia** (Euphorbiaceae)

From the latex of *E. characias*, Barbieri et al. (1983) isolated a Gal-binding lectin on acid-treated agarose which is a dimer consisting of 40,000 MW subunits. In screening the latex of several species of *E.* and the related *Elaeophorbia*, also Lynn and Clevetter-Radford (1986) isolated Gal-binding lectins by affinity chromatography on a commercial lactose-gel, followed by gel filtration. The molecular weights of these lectins range from 60,000 to 67,000, with subunit molecular weights between 27,000 and 38,000.

**Falcata japonica** (Leguminosae)

A lectin directed against the human blood group A was isolated by adsorption to GalNAc-starch as an affinity adsorbent and by elution with an acidic buffer (Nakajima and Furukawa 1979).

### Fomes fomentarius (Poriales)

Echter Zunderschwamm

From the fruiting bodies of this fungus, Hořejší and Kocourek (1978) isolated a lectin by means of an allyl-α-D-galactoside/acrylamide copolymer which was eluted with Gal. The native lectin has a molecular weight of 60,000 and disintegrates into 35,000, 21,000 and 10,000 MW subunits in sodium dodecyl sulfate.

### Galanthus nivalis (Amaryllidaceae)

Snowdrop; Schneeglöckchen; perce-neige

Peuman's group (De Meirsman et al. 1986) isolated a lectin from the bulbs by affinity chromatography on immobilized mannose. The tetrameric lectin (42,000 MW) binds to mannose and melibiose and is composed of four identical subunits.

### Glycine max (Leguminosae)

Soybean; Sojabohne; soja

The seed lectin, called SBA (soybean agglutinin), is one of the lectins studied most thoroughly. In addition to affinity adsorbents from spacer-bound β-galactosyl groups coupled to a matrix (Allen and Neuberger 1975; Gordon at al. 1972) which have been mentioned already and are reviewed in reference (Goldstein and Hayes 1978), Bloch and Burger (1974) introduced an affinity adsorbent from immobilized p-aminophenyl-n-acetyl-β-galactosaminide from which the lectin was desorbed with lactose. Affinity chromatography on immobilized human immunoglobulins was used by Franz and Ziska (1981), and on immobilized glycopeptides from hog gastric mucin by Maylié-Pfenninger and Jamieson (1979). Freier et al. (1985) used a similar but more simple approach. They immobilized desialyzed hog gastric mucin to agarose without prior proteolysis and obtained a high capacity adsorbent from which the lectin was desorbed with Gal, GalNAc or borate.

A lectin virtually identical with the *G. max* lectin was isolated from the wild soybean, *G. soja* (Pueppke et al. 1982).

An agglutinating protein which also acts as an α-galactosidase was isolated by Del Campillo and Shannon (1982) from soybean seeds. They used conventional purification methods though affinity chromatography on melibiose-Sepharose was also effective. This protein is not related to the "classic" seed lectin but rather to non-agglutinating α-galactosidases from other plants. In search of a protein that interacts with exopolysaccharides from soybean-nodulating *Rhizobia*. Rutherford et al. (1986) found a new lectin in *G. max* seeds. It interacts with 4-0-methyl glucuronic acid, a rhizobial exopolysaccharide component.

Affinity chromatography was performed on a partially hydrolyzed methyl-glucuronorhamnan from the *Rhizobia* which does not bind the classic seed lectin. Because of the unspecific ion exchange quality of this adsorbent, further purification steps had to be added. In one of these steps, the glycoprotein nature of the lectin could be utilized by affinity chromatography on Con A-Sepharose.

### Griffonia simplicifolia (*Bandeiraea simplicifolia*, Leguminosae)

The seed lectins of this plant have been reviewed in reference (Goldstein and Hayes 1978). Most of the work on them has been done in the laboratory of I. J. Goldstein. There are two major lectins called lectin I and lectin II. Lectin I has a specificity for α-linked Gal and GalNAc. It consists of two types of subunits which form a series of five isolectins. The mixture of the isolectins I was purified by the use of immobilized melibionic acid. Alternatively, the lectin I mixture was also isolated on guaran copolymerized with acrylamide (Horisberger 1977) or cross-linked with divinylsulfone (Young and Leon 1978) and with melibiose coupled to aminoethyl-polyacryl-amide by reductive amination (Baues and Gray 1977). The isolectin mixture was originally re-solved by the combined use of adsorbents from melibionate and hog mucin and by applying dif-ferent concentrations of the desorbing sugars (Murphy and Goldstein 1978). Later the proce-dure was improved (Delmotte and Goldstein 1978) by the use of an affinity gel to which Gal had been connected via a long spacer arm. The elution was performed with a gradient of stepwise rising methyl-α-galactopyranoside concentrations.

Lectin II consists of only one species. It differs from the isolectins I also in being smaller in molecular weight and in binding a different sugar, GlcNAc. It can be purified on chitin as an af-finity adsorbent (Ebisu and Goldstein 1978), alternatively, GlcNAc bound on a spacer arm to agarose was used (Delmotte and Goldstein 1980). Both lectins bind also to desialyzed hog gas-tric mucin and ovomucoid, with a very high capacity to the latter adsorbent. The lectins I and II can be resolved from each other by starting elution with galactose which desorbs the lectins I, and then changing to borate which removes lectin II from the column (Freier et al. 1985).

Recently, Goldstein's group isolated further lectins from *G. simplicifolia* seeds. Lectin III consti-tutes a series of five GalNAc binding isolectins (Goldstein 1983). Lectin IV was purified on im-mobilized Le[b] blood group substance (Shibata et al. 1982). Though its activity is primarily di-rected against L-fucose, neither the monosaccharide nor simple derivates are able to inhibit the precipitin reaction with a synthetic antigen (Le[b]-bovine serum albumin). In its properties, this lectin differs markedly from lectins I and II. Lectins I, II and IV, but not III are also found in the leaves of *G. simplicifolia* (Lamb et al. 1983).

### Hordeum vulgare (Gramineae)
Barley; Gerste; orge

From the seed embryo, a lectin was isolated by Peumans' group (Peumans et al. 1982b). After extraction of the embryos, heat treatment and neutralization which removed much inactive pro-tein, the purification was achieved by means of a commercial GlcNAc-containing affinity gel. Later, much higher amounts of a lectin indistinguishable from the embryo lectin were found in vegetative parts of the plant as roots, leaves and developing ears (Cammue et al. 1985c). In its physicochemical, biological and immunochemical properties the lectin is very similar to the *Triticum* lectin (Peumans et al. 1982b; Miller and Bowles 1983). This relationship is close enough to allow subunit exchange with the *Triticum* lectin under acidic conditions (Peumans et al. 1982c).

### Hura crepitans (Euphorbiaceae)*
Sandbox tree; Sandbüchsenbaum

The seed extract is known to be highly mitogenic. McPherson and Hoover (1979) isolated a lec-tin from the seeds. After removal of lipids (toluene, chloroform) and extraction with buffer, a column with immobilized blood group active substance from hog intestinal mucosa was used as an affinity adsorbent. Desorption was achieved with lactose. The isolated lectin, however, was not as active in mitosis stimulation as expected from the crude extract.

In contrast, the mitogenic activity increased in the purification protocol of Falasca et al. (1980) which consisted of adsorbtion to acid-treated agarose and elution with Gal. Affinity chromato-graphy on untreated agarose was successfully applied by Pere et al. (1981).

From the latex, Barbieri et al. (1983) and Tovar et al. (1983) isolated several lectins that differ in molecular weight considerably from the seed lectins.

**Laburnum alpinum** (Leguminosae)*

Laburnum, golden chain; Goldregen; cytise

Konami et al. (1983b) described the purification of the seed lectins. They found lectins of two different specificities. The first one, designated lectin I, is specifically inhibited by diacetyl-chitobiose, the second one (lectin II) by Gal. The crude extract was prepurified by salt precipitation and ion exchange, this material was then adsorbed onto a gel which had been prepared by reductive amination of triacetylchitotriose with an amino group substituted agarose. Desorption of lectin I was performed with acetic acid. Lectin II was purified on a galactose containing gel where it was retarded compared to inactive proteins.

Both lectins from *L. alpinum* and from the closely related *L. vulgare* can also be separated by chromatography on immobilized desialyzed hog gastric mucin and by sequential elution with galactose and borate (Freier 1985; Freier et al. 1985). From the bark of *L. anagyroides*, Lutsik and Antonyuk (1982) isolated by conventional methods a lectin with a specificity toward L-fucose and blood group 0(H). The bark lectin undergoes seasonal concentration changes (Antonyuk et al. 1982).

**Lathyrus** (Leguminosae)*

Vetchling; Platterbse; gesse

The seeds of several species contain lectins of the same group as most *Vicia* lectins and the lectins from *Pisum* and *Lens*. They specifically bind mannose and glucose and their α-glycosides and are composed of two pairs of subunits dissimilar in molecular weight ("two-chain lectins"). Generally, the *Lathyrus* lectins can be purified on cross-linked dextrans, e. g. Sephadex (Gupta et al. 1980; Tichá et al. 1980; Kolberg and Sletten 1982; Rougé and Chabert 1983; Rougé and Sousa-Cavada 1984; Sousa-Cavada and Rougé 1985). In several species, isolectins have been found. Though these lectins are very similar to each other and to other two-chain lectins in structure and in sugar specificity, they differ in mitogenic stimulation of human lymphocytes (Borrebaeck and Rougé 1986).

**Lens culinaris** (*L. esculenta*, Leguminosae)*

Lentil; Linse; lentille

The seeds contain two isolectins which belong to the two-chain type. Both are very similar in total and in subunit molecular weight and in sugar specificity. Their isolation by the use of cross-linked dextrans (Sephadex) has already been reviewed in reference (Goldstein and Hayes 1978). The lectins have also been affinity-purified on maltose immobilized by reductive amination (Baues and Gray 1977), on immobilized human immunoglobulins (Franz and Ziska 1981), on a hydroxy group containing polymer which had been glycosylated with glucose or mannose (Filka et al. 1978), and on whole immobilized yeast cells (Ramstorp and Mattiasson 1982).

**Lepidium sativum** (Cruciferae)

Garden cress; Gartenkresse; cresson alénois

Ziska and Franz isolated a homogeneous lectin from the seeds by means of immobilized human immunoglobulin and desorption with a chaotropic reagent, potassium thiocyanate (Franz and Ziska 1981; Ziska and Franz 1982).

Simple monosaccharides did not inhibit at all or only at very high concentrations (GalNAc, N-acetylneuraminic acid) whereas some glycoproteins were quite effective.

44

**Lonchocarpus capassa** (Leguminosae)

Apple-leaf

From the seeds, Joubert et al. (1986) purified a lectin by the use of immobilized Gal. In its sub-unit size and sequence data it is related to other leguminous lectins.

**Lotononis bainesii** (Leguminosae)

Law and Strijdom (1984) isolated lectins from the roots and the seeds. Though from both sources the isolation was achieved by affinity adsorption on similar glycoconjugates, the preparations do not cross-react immunologically and differ in sugar specificity and molecular data. The root lectin is believed to participate in the interaction with *Lotononis*-nodulating *Rhizobia*.

**Lotus tetragonolobus** (*Tetragonolobus purpureus*, Leguminosae)

Asparagus pea; Spargelerbse

A detailed chapter in reference (Goldstein and Hayes 1978) is dedicated to the L-fucose binding lectins of this plant. In more recent work, the lectins have been purified on immobilized p-aminophenyl-L-fucoside (Bloch and Burger 1974), on L-fucose immobilized by the use of di-vinyl sulfone (Allen and Johnson 1977), and on immobilized glycopeptides (Maylié-Pfenninger and Jamieson 1979) or glycoproteins from hog gastric mucin (Freier et al. 1985). In a reexamination of the fucose-binding lectins from *L. tetragonolobus*, Ogata et al. (1985) resolved the fraction which binds to fucose-Sepharose into four individual lectins, three of which do not only bind to L-fucose but also to hog gastric mucin whereas the fourth does not interact with this glycoconjugate.

**Lycopersicon esculentum** (Solanaceae)\*

Tomato; Tomate; tomate

Nachbar et al. (1980; 1982) described the isolation of a lectin from the soluble portion of the fruits which binds to GlcNAc oligosaccharides. As an affinity adsorbent, hog A+H substance insolubilized to polyleucine or alternatively immobilized ovomucoid were used. Elution was performed either with diacetylchitobiose or, in the case of the ovomucoid gel, with an acidic buffer. Kilpatrick (1982) used glutaraldehyde-fixed red blood cells as an affinity adsorbent and desorbed the lectin with a GlcNAc-oligomer mixture. More recently, Kilpatrick et al. (1983) found that the tomato lectin can be purified without affinity chromatography by chromatofocusing taking advantage of its unusually high isoelectric point.

**Maackia amurensis** (Leguminosae)

This plant contains agglutinating and mitogenic proteins. Their isolation has already been reviewed in reference (Goldstein and Hayes 1978).

**Maclura pomifera** (Moraceae)

Osage orange; Osagedorn

The series of affinity adsorbents for the seed lectin which has been reviewed in reference (Goldstein and Hayes 1978) has been complemented by Hořejší and Kocourek (1978) who used the allyl glycoside of GalNAc, copolymerized with acrylamide, by Bausch and Poretz (1977) who introduced immobilized p-aminophenyl-α-D-galactoside, and by Freier et al. (1985) who used desialyzed, immobilized hog gastric mucin as an affinity adsorbent. Similar to the isolectin series from *Phaseolus vulgaris* and *Griffonia simplicifolia*. Bausch et al. (1981) resolved the affinity-purified lectin into five very similar isolectins be means of ion exchange chromatography.

45

**Marasmius oreades** (Tricholomataceae)

Nelkenschwindling

Hořejší and Kocourek (1978) isolated a Gal-binding lectin from the fruiting bodies of this fungus by affinity chromatography on an allyl-α-galactoside/acrylamide copolymer. It consists of 33,000 and 23,000 MW subunits which form an 50,000 MW native molecule.

**Medicago sativa** (Leguminosae)

Alfalfa; Luzerne; luzerne

The seeds contain a lectin which binds to Gal and related sugars. Kamberger (1978) purified the lectin on three different adsorbents: galactosamine coupled to a carboxylated agarose (Allen and Neuberger 1975), GalNAc immobilized by the bisoxirane method (Porath 1974), and agarose which had been partially hydrolyzed (Ersson et al. 1973; Allen and Johnson 1976). Elution was performed with Gal or with an acidic buffer.

**Moluccella laevis** (Labiatae)

Muschelblume

From the seeds, Janssen and Uhlenbruck (1983) separated anti-A and anti-N activities by affinity chromatography on a gel with immobilized GalNAc. The anti-A lectin is composed of two different subunits (MW 28,000 and 30,000).

**Momordica charantia** (Cucurbitacea)*

Bitter pear melon; Balsambirne

Barbieri et al. (1979) isolated a seed lectin according to the protocol of reference (Tomita et al. 1972) on Sepharose 4B. The lectin does not only agglutinate human red blood cells but is an inhibitor of protein synthesis. In contrast to other inhibitors (see *Abrus, Ricinus, Viscum*), this protein is not able to enter the intact cell and is only active in cell-free systems, i. e. in reticulocyte lysates. By the group of Kocourek (Hořejší et al. 1980) the lectin has been isolated using an allyl-α-galactoside/acrylamide copolymer as an affinity adsorbent and eluting with galactose. By conventional methods, Li (1980) resolved two lectins. For both amino acid compositions and N-terminal sequences are given. Majumdar et al. (Majumdar and Surolia 1979) described the isolation on a naturally occurring arabinogalactan, and Ghosh et al. (1981b) used insolubilized guar gum for the isolation of the lectin from *M. dioica*.

**Mucuna flagellipes** (Leguminosae)

Mbadiwe and Agogbua (1978) found an agglutinating activity in the seed extract which was claimed to be specific for the human blood group B. Regrettably, no purification has been attempted. This would have been worthwhile since there are only a few lectins with this specificity.

**Onobrychis viciifolia** (Leguminosae)

Sainfoin; Esparsette; sainfoin

Hapner and Robbins (1979) isolated a seed lectin by the use of mannose immobilized to agarose by the divinylsulfone procedure. The lectin consists of 26,000 MW subunits which combine to dimers. In its sugar specificity and amino acid sequences, it is closely related to other mannose/glucose-binding leguminous lectins (Hapner and Robbins 1979; Kouchalakos and Hapner 1984; Kouchalakos et al. 1984). The lectin has also been isolated on Sephadex (Young et al. 1982).

A lectin indistinguishable from the seed lectin occurs in the roots (Hapner and Robbins 1979).

46

### Ononis spinosa (Leguminosae)

Restharrow; Hauhechel; bugrane

Hořejší and Kocourek (1978) used an allyl-α-D-galactoside/acrylamide copolymer to purify a lectin from the roots. In spite of its sugar specificity, this lectin preferentially agglutinates red cells of blood group 0(H). A similar lectin has been isolated from the roots of *O. hircina* by the same workers (Hořejší et al. 1978a).

### Oryza sativa (Gramineae)

Rice; Reis; riz

A lectin with properties similar to those from wheat germ (WGA, see *Triticum*) has been isolated by several groups. Tsuda (1979) used a combination of conventional procedures and affinity chromatography on immobilized ovomucoid. Though this lectin has about the same total molecular weight as WGA, on reductive dissociation it disintegrates into smaller entities than WGA does. By chromatography on immobilized GlcNAc, Indravathamma and Seshadri (1980) purified the lectin which in their hands displayed molecular properties different from those of reference (Tsuda 1979). Isolation, properties and biosynthesis of the lectin are described by Peumans and Stinissen (1982) and by Stinissen et al. 1982. An improved isolation procedure has been worked out by Poola et al. (1986) who used a spacer-coupled GlcNAc β-thioglycoside as an affinity adsorbent.

### Persea americana (Lauraceae)

Avocado; Avocadobirne; avocatier

From the seeds an agglutinating factor has been enriched, but no detailed molecular data are given. Possibly, the factor is not a protein (Meade et al. 1980; Yaakobovich and Neeman 1983).

### Phaseolus aureus (Leguminosae)

see *Vigna radiata*

### Phaseolus coccineus (Leguminosae)

Scarlet runner; Feuerbohne; haricot d'Espagne

Angelisová and Haskovec (1978) isolated a mitogenic lectin from the seeds by salt precipitation and ion exchange. Though in mitogenic activity and subunit structure this lectin resembles the well-known *P. vulgaris* lectin, its mitogenic activity is inhibited by methyl-α-D-mannoside. By using immobilized red blood cell stroma as an affinity adsorbent and an acidic buffer for desorption, Ochoa and Kristiansen (1978; 1982) isolated the lectins from *P. coccineus* and *P. vulgaris*.

### Phaseolus lunatus limensis (Leguminosae)

Lima bean; Lima-, Mondbohne

The standard procedure for the isolation of the tetrameric and octameric forms of the seed lectin as reviewed in reference (Goldstein and Hayes 1978) consists in affinity chromatography on insolubilized blood group A substance, followed by gel filtration. The specificity of the lectin toward α-linked GalNAc residues is rather high but not as narrow as in *Dolichos biflorus*. Hence then lectin cannot be purified on galactose containing polymers (Allen and Johnson 1976; Lönngren et al. 1976), but is adsorbed onto a gel prepared from galactosamine coupled to a carboxyl containing gel (Freier 1985). The lectin has also been purified on a α-linked GalNAc-containing copolymer (Read and Northcote 1983), and on a gel which contains mucin derived glycopeptides (Maylié-Pfenninger and Jamieson 1979). An adsorbent with a fairly high capacity has been prepared by immobilizing desialylized hog gastric mucin (Freier et al. 1985).

**Phaseolus vulgaris** (Leguminosae)*

Kidney bean and other variants; Gartenbohne; haricot

The seeds contain as well red cell agglutinating as lymphocyte mitosis stimulating activities. Both activities are due to a series of five tetrameric isolectins ($E_4$, $E_3L$, $E_2L_2$, $EL_3$, $L_4$) formed from two subunit types E and L. Isolation of the isolectin mixture on immobilized thyroglobulin or fetuin and subsequent resolution into the individual isolectins by ion exchange and physico-chemical properties have been reviewed in reference (Goldstein and Hayes 1978). The lectins do not react with simple sugars but only with complex oligosaccharide structures. More recent work (Osborn et al. 1983) has shown that there occur more than just two subunit types which results in a more complex isolectin pattern.

The lectins have also been purified by means of immobilized red cell stroma (Ochoa and Kristiansen 1978). Datta and Ray (1979) and Ohtani et al. (1980) report that due to their glycoprotein nature the lectins can be purified on Con A-Sepharose. In our hands, this procedure did not lead to pure preparations (Fleischmann 1985), whereas affinity chromatography on immobilized desialylized hog gastric mucin or, with a better capacity, immobilized ovomucoid (Fleischmann et al. 1985; Freier et al. 1985) leads to highly purified lectin. The latter adsorbent makes it possible to resolve the individual isolectins in one run (Fleischmann et al. 1985). For this purpose, the crude extract is brought from pH 8 to pH 5, centrifuged, the supernatant readjusted to pH 8, and dialyzed against 0.05 M Tris/acetate ph 8.0, containing 1 mM $CaCl_2$ and $MgCl_2$. From the prepurified extract, the lectins are adsorbed to immobilized ovomucoid which had been freshly prepared according to the method of Lineweaver and Murray (1947). The isolectins are then desorbed sequentially with buffers of rising borate concentrations. The isolectin $L_4$ which is the most potent mitogen leaves the column first (at 15 mM borate). It is recommendable to utilize only 50 %, 75 % of the column's capacity since the isolectin $L_4$ is bound less firmly than the others. Therefore, during application of the extract it may be partially lost by displacement by other isolectins.

From the seeds of a *P. vulgaris* variant, Pinto III, previously regarded as lectin-free, Pusztai et al. (1981) by the use of conventional methods isolated a lectin which is active against pronase-treated rat blood cells. It differs from the "classic" lectins by its molecular weight and its subunit composition, but shows a slight immunochemical cross-reactivity against them.

**Phytolacca americana** (Phytolaccaceae)*

Pokeweed; Kermesbeere; phytolacca

Mitogenic lectins have been isolated by the use of conventional and affinity techniques from the roots. Remarkably, one of these lectins is mitogenic for T- and B-cells as well (see Goldstein and Hayes 1978). Lectin mixtures can also be isolated on immobilized hog gastric mucin and ovomucoid (Freier et al. 1985).

**Pinellia ternata** (Araceae)

Park et al. (1981) purified a lectin from the roots by conventional methods. The lectin consists of only one subunit (160,000 MW). It does not react with human, but with several animal red blood cells. No simple monosaccharide was able to inhibit agglutination. Lee et al. (1983) purified the lectin by the use of immobilized thyroglobulin, elution was performed with water and urea.

**Pisum sativum** (Leguminosae)*

Pea; Erbse; pois

The seed lectin belongs to the mannose/glucose binding two-chain lectins. Its isolation on glucose polymers (mostly commercial cross-linked dextran, Sephadex) has been known since long and is reviewed in reference (Goldstein and Hayes 1978). Van Driessche et al. (1978) proposed an improvement which leads to a fourfold yield by performing the whole isolation procedure in the presence of bivalent ions ($MnCl_2$, $CaCl_2$) and at a high salt (NaCl) concentration.

### Psophocarpus tetragonolobus (Leguminosae)
Winged bean; Goabohne

The seeds contain lectins which according to their isoelectric points are classified as acidic and basic lectins. Pueppke (1979) isolated a galactose-binding acidic lectin by affinity chromato-graphy. Out of six adsorbents which he tested, a lactosaminyl gel prepared according to refer-ence (Baues and Gray 1977) was the most effective. Appukuttan and Basu (1981) used an affin-ity gel prepared from a carboxyl group bearing gel aminated with galactosamine. They found only one lectin which was composed of one subunit. Kortt used a combination of conventional methods (Kortt 1984; 1985a) and affinity techniques on gel containing immobilized lactose and melibiose (Kortt 1985b). He found that both the acidic and the basic fraction are composed of several individual lectins. All lectins are very similar in amino acid sequences and in sugar specificity which is directed against GalNAc and Gal. The acidic lectins show a remarkably low affinity for these sugars. Both lectin groups accept α- and β-linked glycosides with nearly equal efficiency. The basic lectin has also been purified by Higuchi and Iwai (1985) on immobilized p-aminophenyl-β-D-galactopyranoside. From the tubers, Shet et al. (1985) isolated by conven-tional methods a lectin which is similar to the seed lectins.

### Ricinus communis (Euphorbiaceae)*
Castor bean; Wunderbaum, Ricinus; ricin

Ricinus seeds have been known since long as being extremely toxic. Since seed extracts also agglutinate red cells, it was assumed that toxin and agglutinin are due to the same protein. Later, however, it turned out that both qualities belong to different but related proteins. The agglutinin (about 120,000 MW) is composed of four subunits, the toxin (about 60,000 MW) of two. Both bind to galactose and related sugars. Much work has been devoted to the toxin with the aim of synthesizing target-directed drugs (see *Abrus*). The toxin is composed of a toxic sub-unit (A, effectomer), which inhibits protein synthesis in eukaryotes, and a Gal-binding subunit (B, haptomer), which is responsible for the attachment to cell surfaces.

Both, agglutinin and toxin, bind to agarose and are desorbed with Gal-containing buffers. Sep-aration can be achieved by gel filtration, ion exchange, or by selective desorption of the toxin from an agarose column by GalNAc. Detailed reviews about the purification and properties of these proteins are found in references (Goldstein and Hayes 1978; Olsnes 1978a) and more re-cently in Olsnes and Pihl (1982).

Lin and Li (1980) used a combination of conventional purification procedures to resolve the agglutinin and toxin isoforms. Most workers, however, confine themselves to the separation of the agglutinins from the toxins.

Apart from native agarose or acid-treated agarose with improved capacity (Allen and Johnson 1976), also other polymers containing galactose are suited as affinity adsorbents, such as cross-linked guaran (Appukuttan and Bachhawat 1979; Lönngren et al. 1976), arabinogalactan (Majumdar and Surolia 1979), a lactosaminyl gel prepared by reductive amination (Baues and Gray 1977), a glycosylated synthetic polymer (Filka et al. 1978), copolymers from allyl-β-lac-toside and acrylamide (Hořejší and Kocourek 1978), immobilized naturally occurring glyco-peptides and glycoproteins (Freier et al. 1985; Lee et al. 1983; Maylié-Pfenninger and Jamieson 1979) and red cell stroma (Genaud et al. 1982).

49

The difference in carbohydrate affinity between the toxin and the lectin can be utilized for their separation. Hořejší (1979b) applied an agglutinin/toxin mixture to an α-L-rhamnosyl polyacrylamide gel. Since the immobilized sugar is similar to β-Gal (deviations only in positions 3 and 6), the agglutinin is adsorbed whereas the toxin is not. Simmons and Russel (1985) used immobilized p-aminophenyl-1-thio-β-galactoside to adsorb both proteins from the crude extract. Separation was achieved with a Gal gradient. Freier et al. (1985) used desialylized immobilized hog gastric mucin, from which the toxin could be desorbed with GalNAc whereas the agglutinin needed Gal for desorption.

Affinity techniques have also been used to isolate the A-subunit of the toxin. Appukuttan and Bachhawat (1979) adsorbed the whole toxin to a guar gel. On treatment of the loaded column with mercaptoethanol, the A-chain was split off and eluted, the remaining B-chain was subsequently desorbed with lactose. In a similar way, Fulton et a. (1986) used acid-treated agarose to adsorb the toxin, the A-chain was then removed by treatment with mercaptoethanol and the B-chain eluted with Gal. Further resolution of the A-chain into two isoforms was achieved by ion exchange. The last traces of contaminating B-chain were removed from the A-chain by chromatography on immobilized asialofetuin and immobilized monoclonal anti-B-chain antibody whereas from the B-chain preparation A-chain contaminants were separated by ion exchange and immobilized anti-A-chain antibody.

### Robinia pseudoacacia (Leguminosae)*
Black locust; Robinie; robinier

The seed lectin had been purified and characterized earlier by conventional techniques (see Goldstein and Hayes 1978). Its specificity is directed towards complex carbohydrates. More recently, the lectin has been purified by adsorption to immobilized fetuin and elution with galactose (McPherson and Hoover 1979) and by adsorption to immobilized desialomucin or ovomucoid and desorption with borate (Freier et al. 1985). Wantyghem et al. (1984; 1986) used conventional techniques to resolve the seed lectins into two fractions which differ in total molecular weight (59,000 and 105,000) and ability to agglutinate red cells and to stimulate lymphocyte mitosis. Apparently the same lectins were also resolved by affinity chromatography on ovomucoid-Sepharose by subsequent elution with borate and an acidic buffer (Fleischmann and Rüdiger 1986). The affinity-purified preparations of reference (Fleischmann and Rüdiger 1986), however, have total molecular weights of 120,000 and 110,000 whereas as far as tested all other properties are similar.

A lectin similar to one of the seed lectins was also isolated from *R. pseudoacacia* bark by selective adsorption to formaldehyde-fixed red blood cells (Hořejší et al. 1978b). Recently the bark lectin was shown to undergo remarkable seasonal changes (Antonyuk 1984; Nsimba-Lubaki and Peumans 1986). The distribution of lectin activity in different tissues of *R. pseudoacacia* was studied by Gietl and Ziegler (1980). An extremely high agglutinating activity was found in the phloem. The sieve tube sap lectin was purified by Gietl et al. (1979). In its subunit composition, it differs from the seed lectin, but the total molecular weight (95,000, 100,000) determined by two methods corresponds to the seed lectin molecular weight.

### Sambucus (Caprifoliaceae)
Elder; Holunder; sureau

From the bark of *S. nigra* (Broekaert et al. 1984), *S. ebulus* and *S. racemosa* (Nsimba-Lubaki et al. 1986), Peumans' group isolated lectins by affinity chromatography on immobilized fetuin. These lectins are composed of 32,000, 37,000 MW-subunits with a total molecular weight of about 140,000. In agreement with leguminous lectins, the *Sambucus* lectins are localized in protein bodies of the phloem parenchyma (Greenwood et al. 1986). As already found for the *Robinia* bark lectin, the *Sambucus* lectin changes its activity seasonally (Greenwood et al. 1986; Nsimba-Lubaki and Peumans 1986).

### Sarothamnus scoparius (Leguminosae)
see *Cytisus scoparius*

### Secale cereale (Gramineae)
Rye; Roggen; seigle
From rye germ Kubánek et al. (1982) purified three isolectins by a combination of affinity chromatography on chitin with conventional techniques. A lectin was also isolated by Peumans et al. (1982b) by affinity chromatography on a commercial GlcNAc gel. The rye lectin resembles the *Triticum* lectin (wheat germ agglutinin, WGA) in molecular characteristics and in binding specificity. It is even able to exchange subunits with WGA and with other cereal lectins.

### Sesamum indicum (Pedaliaceae)
Sesame; Sesam; sésame
The seeds contain a blood group unspecific, GlcNAc binding lectin which was purified using chitin as an affinity adsorbent (Zhu and Sun 1982). The lectin (MW 56,000) is composed of two unequal subunits (30,000 and 26,000) connected by disulfide bonds.

### Solanum tuberosum (Solanaceae)*
Potato; Kartoffel; pomme de terre
Isolation and properties of the potato tuber lectin have already been reviewed in reference (Goldstein and Hayes 1978). The lectin binds to O-linked oligomers of GlcNAc and is unspecific in the AB0(H) blood group system. It greatly differs from the leguminous lectins in containing about 50% sugar, with a preponderance of arabinose which is linked to the protein O-glycosidically at hydroxyproline residues.
In addition to conventional or affinity purifications reviewed in reference (Allen and Neuberger 1978), the lectin has also been purified by adsorption to immobilized N, N', N''-triacetylchitotriose (Desai and Allen 1979; Matsumoto et al. 1983) and elution with acetic acid, and by the use of immobilized thyroglobulin (Lee et al. 1983) or fetuin (Owen and Northcote 1980) as affinity adsorbents. In the latter case, elution had to be performed with N, N', N''-chitotriose because acidic buffers were ineffective.
McCurrach and Kilpatrick (1986) purified the lectin to homogeneity by chromatofocusing and gel filtration without employing affinity chromatography. In the same paper, the authors also isolated the lectin from potato fruits and showed it to be similar to, but not identical with, the tuber lectin. The tuber lectin had earlier been isolated by the same group by adsorption to glutaraldehyde-fixed red blood cells (Kilpatrick 1980). Chitin was suited as affinity adsorbent for a lectin from the fruits of *S. nigrum* (Colceag 1985).

### Sophora japonica (Leguminosae)*
Japanese pagoda tree; Schnurbaum
Affinity adsorption of the seed lectin to insolubilized hog gastric mucin and desorption with Gal have been reviewed in reference (Goldstein and Hayes 1978). The lectin can also be purified by the use of acid-treated agarose but not on native one (Allen and Johnson 1976), and on desialylized, immobilized hog gastric mucin (Freier et al. 1985).

**Trichosanthes kirilowii** (Cucurbitaceae)

Haarblume

Lee et al. isolated a lectin from the seeds by adsorption to acid-treated agarose and desorption with Gal (Wang et al. 1983) and characterized its sugar specificity (Lee and Wang 1983). The lectin could also be isolated on immobilized thyroglobulin (Lee et al. 1983).

**Trifolium repens** (Leguminosae)

White clover; Weißklee; trèfle blanc

One of the possible biological functions of lectins frequently discussed is their interaction with nodulating *Rhizobia* (for a review see Rüdiger 1984). Dazzo et al. (1978) in search of a factor that is able to interact with clover-nodulating *Rhizobia*, purified a lectin from the seeds by conventional methods. As an analytical method, chromatographic fractions were screened for their ability to agglutinate *Rhizobium trifolii* cells. More important in view of the biological function was the finding that proteins could be washed from young seedling roots with a 2-deoxyglucose containing buffer. The major fraction of this protein proved to be identical with the seed lectin in several criteria (electrophoretical behavior, immunochemical cross-reactivity, agglutination of *R. trifolii* and binding to root hairs). Though the lectins from both locations are identical, it is not the seed lectin that migrates to the roots during germination, but the interaction with Rhizobia is mediated by lectin that is newly synthesized in the root cells (Truchet et al. 1986).

**Triticum vulgare** (*T. aestivum, T. sativum*, Gramineae)\*

Wheat; Weizen, blé

The seed lectin is localized in the germ and is therefore mostly referred to as wheat germ agglutinin, WGA. It has been studied intensively and rather early and is reviewed thoroughly in reference (Goldstein and Hayes 1978). Most studies were motivated by the observation that WGA agglutinates transformed cells to a greater extent than the normal parent cells. Purification protocols make use of affinity chromatography either on synthetic β-linked GlcNAc-containing adsorbents, immobilized ovomucoid, or chitin. In addition to the purification procedures cited in reference (Goldstein and Hayes 1978). WGA has also been purified by affinity chromatography on immobilized p-aminophenyl-β-GlcNAc (Bloch and Burger 1974) and on GlcNAc (Vretblad 1976) or GlcNAc oligomers (Kumar et al. 1982) immobilized to agarose by the bisoxirane method. The latter adsorbent has a high capacity. High capacities are also achieved by the reductive amination methods (Baues and Gray 1977; Matsumoto et al. 1981). Schnaar and Lee (1975) built up an affinity adsorbent from polymeric acrylic acid bearing active ester groups. These were replaced aminolytically by aminohexyl glycosides. By using the β-GlcNAc derivative, the authors obtained a resin which was able to adsorb WGA and to release it on lowering the pH. With a pH gradient partial resolution into isolectins could be achieved. Chiang et al. (1979) synthesized aminohexyl-glycosides and coupled them to agarose by the cyanogen bromide procedure. By the use of the β-GlcNAc derivative an adsorbent suited for the isolation of WGA was obtained.

Lutsik (1984) prepared glutaraldehyde cross-linked ovomucoid and applied it to the purification of several lectins, with the best result for WGA. Franz and Ziska (1981) isolated the lectin on immobilized human immunoglobulins, Maylié-Pfenninger and Jamieson (1979) prepared, isolated and immobilized glycopeptides from hog gastric mucin for the isolation WGA and other lectins, and Freier et al. (1985) used desialylized, immobilized hog gastric mucin as an affinity adsorbent for several lectins, among them WGA.

In contrast to most leguminous lectins, WGA is devoid of carbohydrate and contains great amounts of cystine. It is composed of two subunits (MW 18,000) making up a 36,000 MW molecule. In ion exchange, WGA can be split up into several isolectins. Under proper conditions, WGA is able to exchange subunits as well with other WGA molecules as with related lectins from other Gramineae (Peumans et al. 1982c).

Köttgen et al. (1982) found that wheat gluten possesses lectin-like properties. The authors regard this to be the reason for the well-known gluten sensitivity in coeliac disease. Recently Kolberg and Sollid (1985) isolated a lectin from gluten and showed it to be identical with WGA. Further work is necessary to settle the question whether coeliac disease is caused by a further, as far unknown wheat lectin or by residual WGA in the usual gluten preparations.

### Tulipa gesneriana (Liliaceae)

Tulip; Tulpe; tulipe

From the bulbs, Oda and Minami (1986) isolated a lectin which agglutinates yeast cells but no erythrocytes. Purification was achieved by affinity chromatography on immobilized mannan. In contrast to mannose-binding leguminous lectins, the tulip lectin does not recognize glucose.

### Ulex europaeus (Leguminosae)*

Furze, gorse; Stechginster; ajonc

In the seeds two lectins of different specificities occur. The quantity of both is extremely low compared with other leguminous lectins (Freier et al. 1985). The first one to be detected, lectin I, specifically agglutinates red cells of blood group 0(H) and is inhibited by fucose. It is frequently used for blood group typing and as a cell probe. Lectin II, though moderately preferring blood group 0(H) to A and B, is not inhibited by the terminal sugar of this blood group but by oligomeric GlcNAc.

Lectin I has been purified by affinity chromatography on fucosyl containing gels as reviewed in reference (Goldstein and Hayes 1978). In more recent work, lectin I has also been purified on fucose immobilized to agarose by divinylsulfone (Allen and Johnson 1977), on an adsorbent from a synthetical polymer rich in hydroxy groups which had been glycosylated with L-fucose (Richardson 1984), on immobilized human immunoglobulins (Franz and Ziska 1981), on immobilized glycopeptides derived from hog gastric mucin (Maylié-Pfenninger and Jamieson 1979), and on immobilized desialylized hog gastric mucin (Freier et al. 1985). In any case, desorption was accomplished by the hapten sugar, L-fucose. The method of reference (Freier et al. 1985) permits to resolve lectin I from lectin II by elution of the mucin gel first with L-fucose, then with borate.

For the isolation of the lectin II, Hořejší (1979a) used an affinity adsorbent obtained by copolymerization of maleylated hog gastric pepton and acrylamide. The lectin was desorbed with acid. Pereira et al. (1979) adsorbed the lectin II in a batchwise procedure to polyleucyl hog blood group substance and desorbed it with ethylene glycol in form of a crude preparation which still contained lectin I. It was refined on a commercial fucosyl-gel which retained lectin I, whereas lectin II passed through. Konami et al. (1981) prepurified the extract on an ion exchange column in order to remove the N-acetylhexosaminidase activity which otherwise would destroy the affinity adsorbent. The lectin was then applied to a chitobiose-agarose column and desorbed with acetic acid.

### Urtica dioica (Urticaceae)

Stinging nettle; Brennessel; ortie

From the rhizomes Peumans et al. (1984) isolated a lectin. The essential step was adsorption to chitin and desorption with acid. The lectin is unusually small (MW 8,500) and agglutinates very weakly and unspecifically. Agglutination is not inhibited by GlcNAc, but by its oligomers. The

N-terminal portion of its sequence turned out to be significantly homologous to WGA from *Triticum* (Chapot et al. 1986).

### Vicia bungei (Leguminosae)
Zheng et al. (1985) isolated a seed lectin by affinity chromatography on Sephadex or on immobilized thyroglobulin. It binds to mannose and glucose but differs from the usual Vicieae two-chain lectins in its molecular weight.

### Vicia cracca (Leguminosae)*
Common vetch; Vogelwicke
The seeds contain lectins of two different specificities (see also Goldstein and Hayes 1978). The first one belongs to the two-chain type also known from other members of the Vicieae subtribe, as *Pisum, Lens, Lathyrus*, and other *Vicia* species. It can therefore be purified by the use of Sephadex and other α-linked glucose polymers (Baumann et al. 1982). The amino acid sequence of its smaller subunit is highly homologous with other Vicieae lectins (Baumann et al. 1982).

The second lectin has been known since long because of its specificity for the human blood group A and for α-linked GalNAc (Goldstein and Hayes 1978). It has been isolated and resolved into two groups of isolectins by applying a pH-gradient to an affinity gel (Rüdiger 1977) prepared from galactosamine and a carboxyl group containing matrix (Allen and Neuberger 1975). The isolectin mixture has also been purified by affinity chromatography on a copolymer of allyl-α-GalNAc and acrylamide (Hořejší and Kocourek 1978), on cross-linked hog gastric pepton (Karhi and Gahmberg 1980), and on immobilized desialylized hog gastric mucin (Freier et al. 1985). The latter method permits to separate the lectins of both specificities in one run. The GalNAc-binding lectin belongs to the one-chain type and consists of four subunits equal in molecular weight. Karhi et al. (1980) give molecular weight data somewhat lower than those of references (Rüdiger 1977 and Allen 1979). The lectin has been shown to inhibit in vitro protein synthesis in a cell-free system after treatment with mercaptoethanol (Barbieri et al. 1979), which is surprising in view of the fact that it does not contain cysteine or methionine (Hořejší and Kocourek 1978). Both lectins interact with each other in vitro (Baumann and Rüdiger 1981) which may be important in terms of their biological function at their common in vivo site, the cotyledon protein bodies. In the amino acid sequence of the N-terminal portion, the Gal-NAc-specific lectin shows homology with the larger chain of the Man/Glc-specific lectin, but it is not as closely related to the latter one as Man/Glc-specific lectins from other Vicieae (Baumann et al. 1979).

### Vicia ervilea (Leguminosae)
Linsenwicke
As reviewed in reference (Goldstein and Hayes 1978), the seeds contain a mannose/glucose-binding lectin with the architecture of a two-chain *Vicieae* lectin.

### Vicia faba (Leguminosae)*
Broad bean; Saubohne; fève de marais
The seed lectin belongs to the mannose/glucose-binding two-chain lectins. The most simple means to purify it is affinity chromatography on cross-linked dextrans. Due to their low capacity, relatively large columns have to be used. On a synthetic adsorbent, 3-0-methylglucosamine coupled to a carboxyl resin, about tenfold higher capacities are achieved (Allen et al. 1978). Purification and properties of the lectin have been reviewed in reference (Goldstein and Hayes

1978). The lectin can also be purified on immobilized naturally occurring glycoproteins (Freier et al. 1985; Lee et al. 1983). These adsorbents, however, have low capacities. Next to Con A, the *V.faba* lectin was the second lectin whose complete amino acid sequence has been determined (Hemperly et al. 1979; Hopp et al. 1982). Surprisingly, if its subunits are aligned to the Con A sequence written in a circle, they are highly homologous but are circularly permutated, i. e., the Con A sequence bridges the gap between both *V.faba* lectin subunits and vice versa. Later, this result has been confirmed with several other lectins (e. g., from lentil and soybean). An explanation of this unusual relation may be the findings of Carrington et al. (1985), who proposed a posttranslational rearrangement. Of the *V.faba* lectin, the precursor which contains both chains plus an additional N-terminal signal peptide has been identified (Hemperly et al. 1982).

Datta et al. (1984) isolated a lectin from seeds by affinity chromatography on chitin. Its specificity for β-linked GlcNAc residues suggests that it is different from the "classic" lectin, in spite of the similarity between both in terms of their molecular structure.

Dey et al. (1982a; 1982b; 1986) purified multiple forms of α-galactosidases and showed them to possess not only enzymatical but also a lectin activity which is directed against mannose and glucose. Malý et al. (1985) measured the interaction of these enzyme-lectins by using the technique of affinity electrophoresis.

From the leaves of *V.faba*, Ito (1986) isolated a lectin activity by affinity chromatography on immobilized heparin. Agglutination is inhibited only by dissolved heparin, but not by the other carbohydrates tested including polyanionic polysaccharides. So far, no molecular data are given.

### Vicia graminea (Leguminosae)
The seeds contain a lectin the chemical and binding properties of which have been reviewed in reference (Goldstein and Hayes 1978). The lectin is directed against the rare blood group N. It is not inhibited by usual mono- or oligosaccharides but only by glycoproteins from N-positive cells. It has been purified by conventional techniques. In its chemical properties, it fits into the usual frame of leguminous lectins. More recently Lisowska et al. (1976; Duk and Lisowska 1981) purified the lectin by means of immobilized desialyzed glycoproteins from human red cells of group 0(H). Prigent et al. (1982) used as an essential step the adsorption of the lectin to immobilized Con A and desorption with α-methyl-D-mannoside.

### Vicia hirsuta (Leguminosae)
Hairy vetch; Zitterwicke
Solheim (1983) isolated a series of isolectins from the seeds by affinity chromatography using Sephadex. The major lectin which was characterized more closely resembles in sugar specificity, subunit and amino acid composition other mannose/glucose-binding Vicieae lectins.

### Vicia sativa (Leguminosae)
Futter-, Saatwicke
By affinity chromatography using Sephadex, a seed lectin which binds to mannose and glucose has been isolated (Falasca et al. 1979; Rossi et al. 1979; Gebauer et al. 1979). The lectin belongs to the two-chain Vicieae group to which it is closely related also by its amino acid sequence (Gebauer et al. 1981).

### Vicia tetrasperma (Leguminosae)
Fadenwicke
From the seeds, Kochibe (1986) isolated lectins by affinity chromatography on Sephadex,

which in their sugar and immunochemical specificity, but not in their molecular architecture, are similar to other mannose/glucose-binding Vicieae lectins.

### Vicia unijuga (Leguminosae)
In the leaves, an anti-N-lectin activity has been found (Tajima et al. 1977). The lectin has been purified by precipitation and chromatography on Sephadex (Lutsik et al. 1977).

### Vicia villosa (Leguminosae)
Hairy vetch; Zottelwicke
The seeds contain lectins which bind to α-linked GalNAc. They have attracted considerable attention since they selectively bind to a surface component unique to murine cytotoxic lymphocytes. The lectins have been purified on human immobilized blood group A by Kimura et al. (1979), on a synthetic polyhydroxy polymer glycosylated with GalNAc, and on galactosamine coupled to a carboxyl group containing matrix (Grubhoffer et al. 1981). By applying the crude material to immobilized porcine blood group substance and eluting with a stepwise gradient of rising GalNAc concentrations, the lectins could be resolved into three isolectins ($A_4$, $A_2B_2$, and $B_4$) composed of two subunits MW 35,900 and 33,600 (Tollefsen and Kornfeld 1983).

### Vigna radiata (*Phaseolus aureus*, Leguminosae)
Mung bean; Mungbohne
Shannon's group (Hankins and Shannon 1978; del Campillo et al. 1981) observed that trypsinized rabbit red blood cells are readily agglutinated by seed extracts but that after a short time the cell layer disaggregates again ("clot-dissolving activity"). By a combination of conventional procedures, they isolated a seed lectin which at the same time is an α-galactosidase. In all steps of the purification, both activities copurify. The protein has a greater subunit and total MW than most other leguminous lectins but is immunochemically related to some of them. The relationship to the α-galactosidase/lectin from *Glycine max* (del Campillo and Shannon 1982) appears to be very close. The protein occurs in a tetrameric form where it possesses both hemagglutinating and enzymatic activity, and in a monomeric form which acts only as an enzyme.
In contrast to the lectin/enzyme isolated by Dey et al. (1982) from *Vicia faba*, the *V. radiata* protein exhibits the same sugar specificity for its enzymatical and its agglutinating activities.
A very similar protein has been isolated by Haas et al. (1981) from *V. radiata* hypocotyl cell walls.

### Vigna unguiculata (Leguminosae)
Cowpea
Robertson and Strength (1983) isolated by conventional methods a seed lectin which agglutinates red cells irrespective of blood group or species. Agglutination is inhibited by an exraordinarily broad panel of monosaccharides.

### Viscum album (Viscaceae, formerly: Loranthaceae)*
Mistletoe; Mistel; gui
A lectin is present in extracts from whole plants, which apparently is the toxic principle of this plant. Mistletoe extracts have been used since long against various diseases including cancer; the properties of the lectin have therefore been studied by several groups.
Luther et al. (1980) used affinity chromatography on glutaraldehyde-treated human B-erythrocytes and elution with Gal. Franz et al. (Franz and Ziska 1981; Franz et al. 1981; Franz et al. 1982) resolved the toxic principle into three fractions. Lectin I binds to partially hydrolyzed

56

agarose, lectin II to immobilized human IgG. Both, lectins I and II, were desorbed with Gal from the respective affinity adsorbent. A third fraction, lectin III, could be adsorbed to immobilized IgG and to galactosamine coupled to a carboxyl group containing matrix but was desorbed only under very harsh conditions (pH 2.6). Acid-treated agarose was also used by Olsnes et al. (1982b). The lectins consist of different kinds of subunits of about 30,000 MW which combine to dimers (lectins II and III) or tetramers (lectin I). Similar to the *Abrus* and *Ricinus* toxins, the whole lectins consist of a toxic subunit (A) which inhibits protein synthesis, and a carbohydrate binding subunit (B). By subsequent treatment of agarose-bound lectin with β-mercaptoethanol and galactose, the A- and B-chain have been separated from each other.

### Wisteria floribunda (Leguminosae)

Wisteria; Glycinie; glycine

The seeds contain a lectin which binds to GalNAc, and a lymphocyte mitogen [see reference Goldstein and Hayes (1978)].

Cheung et al. (1979) isolated the lectin by affinity adsorption to polyleucyl hog gastric mucin and desorption with lactose. The lectin can also be adsorbed to acid-treated agarose (Allen and Johnson 1976) and to desialylized immobilized hog gastric mucin from which it is desorbed with Gal, GalNAc, or borate (Freier et al. 1985). The lectin is a tetramer of identical subunits of 29,000 MW. From *W. sinensis*, an almost identical lectin has been isolated (Freier et al. 1985). From the lectin-free extract, a series of mitogens has been isolated by a combination of ion exchange and preparative isoelectric focusing (Kaladas and Poretz 1979). One of the mitogens, composed of two identical subunits of 32,000 MW, gives optimal stimulation at very low concentrations (0.05 µg/ml). *Wisteria* lectin and mitogens stimulate also mitosis in lymphocytes from thymus-deficient (nu/nu) mice (Freier 1985).

### Zea mays (Gramineae)

Maize, corn; Mais; maïs

Woods et al. (1979) found a factor that agglutinates Bacteria (several *Erwinia* species). By acid and organic solvent precipitation, the factor was enriched. Electrophoretically, two subunits (MW 15,000 and 30,000) were found. Since the factor is heat- and protease-labile and can be inhibited by *Erwinia* lipopolysaccharides or their hydrolyzates, it appears to be a lectin.

*Acknowledgements*

The author wants to express his gratitude to the members of his group for their engagement in experimental work and for critically reading this manuscript.

Thank is also due to the Deutsche Forschungsgemeinschaft and to the Fonds der Chemischen Industrie who supported the work cited.

## 2.4    References

Agrawal BBL, Goldstein IJ (1967) Protein-carbohydrate interaction. VI. Isolation of concanavalin A by specific adsorption on cross-linked dextran gels. Biochim Biophys Acta 147:262–271

Ahmad N, Bansal R, Ahmad A, Rastogi AK, Kidwai JR (1984) Purification of hemagglutinins from *Agaricus bisporus* by affinity chromatography. Indian J Biochem Biophys 21:237–240 cited after Chem Abstr 101:168631j

Allen AK (1979) A lectin from the exudate of the fruit of the vegetable marrow *(Cucurbita pepo)* that has a specificity for β-1.4-linked N-acetylglucosamine oligosaccharides. Biochem J 183:133–137

Allen AK, Neuberger A (1975) A simple method for the preparation of an affinity adsorbent for soybean agglutinin using galactosamine and CH-Sepharose. FEBS Letters 50:362–364

Allen AK, Neuberger A (1978) Potato lectin. Methods Enzymol 50:340–345

Allen AK, Desai NN, Neuberger A (1978) *Vicia faba* lectin. Methods Enzymol 50:335–339

Allen HJ, Johnson EAZ (1976) The isolation of lectins on acid treated agarose. Carbohydr Res 50:121–131

Allen HJ, Johnson EAZ (1977) A simple procedure for the isolation of L-fucose binding lectins from *Ulex europaeus* and *Lotus tetragonolobus*. Carbohydr Res 58:253–265

Angelisová P, Hasokovec C (1978) Isolation and chemical characterization of a highly purified phytomitogen from *Phaseolus coccineus* seeds. Eur J Biochem 83:163–168

Antonyuk VA (1984) Seasonal changes of activity and carbohydrate specificity of lectins in extracts of different organs of *Robinia pseudocacia*. Farm Zh (Kiev) 1984 (6) 47–50 cited after Chem Abstr 102:75798z

Antonyuk VA, Lutsik MD, Ladnaya LYa (1982) Seasonal variations in hemagglutination titers and affinity to carbohydrates of plant extracts containing fucose-specific lectins. Fiziol Rast 29:1219–1224 cited after Chem Abstr 98:505 11m

Appukuttan PS, Bachhawat BK (1979) Separation of polypeptide chains of ricin and the interaction of the A-chain with Cibacron blue $F_3GA$. Biochim Biophys Acta 580:10–14

Appukuttan PS, Basu D (1981) Isolation of an N-acetylgalactosamine-binding protein from winged bean *(Psophocarpus tetragonolobus)*. Anal Biochem 113:253–255

Asryants RA, Duszenkova IV, Nagradova NK (1985) Determination of Sepharose-bound protein with Coomassie Brilliant Blue G-250. Anal Biochem 151:571–574

Banerjee KK, Sen A (1981) Purification and properties of a lectin from the seeds of *Croton tiglium* with hemolytic activity toward rabbit red cells. Arch Biochem Biophys 212:740–753

Barbieri L, Lorenzoni E, Stirpe F (1979) Inhibition of protein synthesis in vitro by a lectin from *Momordica charantia* and by other hemagglutinins. Biochem J 182:633–635

Barbieri L, Zamboni M, Montanaro L, Sperti S, Stirpe F (1980) Purification and properties of different forms of modeccin, the toxin from *Adenia digitata*. Separation of subunits with inhibiting and lectin activity. Biochem J 185:203–210

Barbieri L, Falasca A, Franceschi C, Licastro F, Rossi CA, Stirpe F (1983) Purification and properties of two lectins from the latex of the euphorbiaceous plants *Hura crepitans* (sand-box tree) and *Euphorbia characias* (Mediterranean spurge). Biochem J 215:433–439

Barbieri, L, Falasca AI, Stirpe F (1984) Volkensin, the toxin from *Adenia volkensii* (kilyambiti plant). FEBS Letters 171:277–279

Baues, RJ, Gray GR (1977) Lectin purification on affinity columns containing reductively aminated disaccharides. J Biol Chem 252:57–60

Baumann CM, Rüdiger H (1981) Interactions between the two lectins from *Vicia cracca*. FEBS Letters 136:279–283

Baumann CM, Rüdiger H, Strosberg AD (1979) A comparison of the two lectins from *Vicia cracca*. FEBS Letters 102:216–218

Baumann CM, Strosberg AD, Rüdiger H (1982) Purification and characterization of a mannose/glucose-specific lectin from *Vicia cracca*. Eur J Biochem 122:105–110

Bausch JN, Poretz RD (1977) Purification and properties of the hemagglutinins from *Maclura pomifera* seeds. Biochemistry 16:5790–5794

Bausch JN, Richey J, Poretz RD (1981) Five structurally related proteins from affinity-purified *Maclura pomifera* lectin. Biochemistry 20:2618–2620

Blacik LJ, Breen M, Weinstein HG, Sitting RA, Cole M (1972) An anti-$A_1$ lectin in the seeds of *Amphicarpaea bracteata*. Biochim Biophys Acta 538:225–231

Bloch R, Burger MM (1974) A rapid procedure for derivatizing agarose with a variety of carbohydrates: its use for affinity chromatography of lectins. FEBS Letters 44:286–289

Borrebaeck CAK, Rougé P (1986) Mitogenic properties of structurally related *Lathyrus* lectins. Arch Biochem Biophys 248:30–34

Broekaert WF, Nsimba-Lubaki M, Peeters B, Peumans WJ (1984) A lectin from elder *(Sambucus nigra)* bark. Biochem J 221:163–169

Cammue BP, Peeters B, Peumans WJ (1985a) Isolation and partial characterization of an N-acetylgalac-

58

tosamine specific lectin from winter-aconite *(Eranthis hyemalis)* root tubers. Biochem J 227:949−955

Cammue BP, Stinissen HM, Peumans WJ (1985b) A new type of cereal lectin from leaves of couch grass *(Agropyrum repens)*. Eur J Biochem 148:315−322

Cammue BP, Stinissen HM, Peumans WJ (1985c) Lectin in vegetative tissues of adult barley plants grown under field conditions. Plant Physiol 78:384−387

Campbell WP, Wrigley CW, Margolis J (1983) Electrophoresis of small proteins in highly concentrated and cross-linked polyacrylamide gradien gels. Anal Biochem 129:31−36

Carrington DM, Auffret A, Hanke DE (1985) Polypeptide ligation occurs during post-translational modification of concanavalin A. Nature (London) 313:64−67

Chapot M-P, Peumans WJ, Strosberg AD (1986) Extensive homologies between lectins from non-leguminous plants. FEBS Letters 195:231−234

Cheung G, Haratz A, Katar M, Skrokov R, Poretz RD (1979) Purification and properties of the hemagglutinin from *Wisteria floribunda* seeds. Biochemistry 18:1646−1650

Chiang CK, McAndrew M, Barker R (1979) 6-Aminohexyl glycopyranosides as ligands for the preparation of affinity adsorbents for the purification of carbohydrate-binding proteins. Carbohydr Res 70:93−102

Colceag J (1985) Purification and some properties of a lectin from *Solanum nigrum* (Solanaceae). Rev. Roum Biochim 22:101−106 cited after Chem Abstr 103:120024c

Crowley JF, Goldstein IJ (1981) *Datura stramonium* lectin: isolation and characterization of the homogeneous lectin. FEBS Letters 130:149−152

Crowley JF, Goldstein IJ (1982) *Datura stramonium* lectin. Methods Enzymol 83:368−373

Cunningham BA, Wang JL, Pflumm MN, Edelman GM (1972) Isolation and proteolytic cleavage of the intact subunit of concanavalin A. Biochemistry 11:3233−3239

Datta TK, Basu PS (1981) Identification, isolation and some properties of lectin from the seeds of Indian coral tree *(Erythrina variegata orientalis)*. Biochem J 197:751−753

Datta TK, Ray MK (1979) Isolation of *Phaseolus vulgaris* lectin (PHA) by affinity chromatography on Concanavalin A-Sepharose 4B gel. Indian J Exp Biol 17:323−324

Datta PK, Basu PS, Datta TK (1984) Isolation and characterization of *Vicia faba* lectin affinity purified on chitin column. Prep Biochem 14:373−387

Dazzo FP, Yanke WE, Brill WJ (1978) Trifoliin: a rhizobium recognition protein from white clover. Biochim Biophys Acta 539:276−286

Del Campillo E, Shannon LM (1982) An α-galactosidase with hemagglutinating properties from soybean seeds. Plant Physiol 69:628−631

Del Campillo E, Shannon LM, Hankins CN (1981) Molecular properties of the enzymic phytohemagglutinin of mung bean. J Biol Chem 256:7177−7180

Delmotte FM, Goldstein IJ (1980) Improved procedures for purification of the *Bandeiraea simplicifolia* I isolectins and *Bandeiraea simplicifolia* II lectin by affinity chromatography. Eur J Biochem 112:219−223

De Meirsman CEP, Nsimba-Lubaki M, Peumans WJ (1986) Melibiose/mannose-specific lectin from snowdrop *(Galanthus nivalis)* bulbs. In: Bøg-Hansen TC, van Driessche E (eds) Lectins − Biology, Biochemistry, Clinical Biochemistry. Vol 5, de Gruyter, Berlin (W), pp 117−123

Desai NN, Allen AK (1979) The purification of potato lectin by affinity chromatography on an N-, N'-, N''-triacetylchitotriose-Sepharose matrix. Anal Biochem 93:88−90

Desai NN, Allen AK, Neuberger A (1981) Some properties of the lectin from *Datura stramonium* (thornapple) and the nature of its glycoprotein linkages. Biochem J 197:345−353

Dean PDG, Johnson WS, Middle FA (eds) (1985) Affinity chromatography; a practical approach. IRL Press Oxford

Dey PM, Naik S, Pridham JB (1982a) The lectin nature of α-galactosidases from *Vicia faba* seeds. FEBS Letters 150:233−237

Dey PM, Pridham JB, Sumar N (1982b) Multiple forms of *Vicia faba* α-galactosidases and their relationships. Phytochemistry 21:2195−2199

Dey PM, Surbhi N, Pridham, JB (1986) *Vicia faba* α-galactosidase with lectin activity. Phytochemistry 25:1057−1061

Duk M, Lisowska E (1981) *Vicia graminea* anti-N lectin: Partial characterization of the purified lectin and its binding to erythrocytes. Eur J Biochem 118:131−136

Ebisu S, Goldstein IJ (1978) *Bandeiraea simplicifolia* lectin II. Methods Enzymol 50:350–354

Eifler R, Ziska P (1980) The lectins from *Agaricus edulis*. Isolation and characterization. Experientia 36:1285–1286

Ersson B, Asperg K, Porath J (1973) The phytohemagglutinin from sunn hemp seeds *(Crotalaria juncea)*. Purification by biospecific affinity chromatography. Biochim Biophys Acta 310:446–452

Etzler ME (1983) Distribution and properties of the *Dolichos biflorus* lectins: a model system for exploring the role of lectins in plants. In: Goldstein IJ, Etzler ME (eds) Chemical Taxonomy, Molecular Biology, and Function of Plant Lectins. Alan R Liss Inc, New York, pp 197–207

Falasca A, Franceschi C, Rossi CA, Stirpe F (1979) Purification and partial characterization of a mitogenic lectin from *Vicia sativa*. Biochim Biophys Acta 577:71–81

Falasca A, Franceschi C, Rossi CA, Stirpe F (1980) Mitogenic and haemagglutinating properties of a lectin purified from *Hura crepitans* seeds. Biochim Biophys Acta 632:95–105

Filka K, Čoupek J, Kocourek J (1978) Studies on lectins. XL. 0-Glycosyl derivatives of Spheron in affinity chromatography of lectins. Biochim Biophys Acta 539:518–528

Fleischmann G (1985) Lektine der Leguminosen und deren pflanzeneigene Bindungspartner – ein Ansatz zum Verständnis ihrer biologischen Funktion. Thesis University of Würzburg

Fleischmann G, Rüdiger H (1986) Isolation, resolution and partial characterization of two *Robinia pseudoacacia* seed lectins. Biol Chem Hoppe-Seyler 367:27–32

Fleischmann G, Mauder I, Illert W, Rüdiger H (1985) A one-step procedure for isolation and resolution of the *Phaseolus vulgaris* isolectins by affinity chromatography. Biol Chem Hoppe-Seyler 366:1029–1032

Franz H, Ziska P (1981) Isolation of lectins with different specificities using immobilized immunoglobulins. In: Bøg-Hansen TC (ed) Lectins – Biology, Biochemistry, Clinical Biochemistry. Vol 1, de Gruyter, Berlin (W) pp 179–184

Franz H, Ziska P, Kindt A (1981) Isolation and properties of three lectins from mistletoe *(Viscum album)*. Biochem J 195:481–484

Franz H, Kindt A, Ziska P, Bielka H, Benndorf R, Venker L (1982) The toxic A-chain of mistletoe lectin I: Isolation and its effect on cell-free protein synthesis. Acta biol med germ 41:K9–K16

Freier T (1985) Isolierung von pflanzlichen Lectinen und ihre Wechselwirkungen mit isologen Samenproteinen. Thesis University of Würzburg

Freier T, Fleischmann G, Rüdiger H (1985) Affinity chromatography on immobilized hog gastric mucin and ovomucoid; a general method for the isolation of lectins. Biol Chem Hoppe-Seyler 366:1023–1028

Fu LF, Chen CC (1982) Isolation of lectins from *Croton tiglium* seeds. Sheng Wu K'o Hsueh 19:45–50 cit after Chem Abstr 97:179970u

Fukuda N, Takazato H, Chinen I, Yomo H (1984) Purification and some properties of deigo *(Erythrina variegata orientalis)* seed lectin. Nippon Nogei Kagaku Kaishi 58:793–8 cited after Chem Abstr 101:167189j

Fulton RJ, Blakey DC, Knowles PP, Uhr JW, Thorpe PE, Vitetta ES (1986) Purification of ricin $A_1$-, $A_2$- and B-chains and characterization of their toxicity. J Biol Chem 261:5314–5319

Gebauer G, Schilz E, Schimpl A, Rüdiger H (1979) Purification and characterization of a mitogenic lectin and a lectin-binding protein from *Vicia sativa*. Hoppe-Seyler's Z Physiol Chem 360:1727–1735

Gebauer G, Schilz E, Rüdiger H (1981) The amino-acid sequence of the α-subunit of the mitogenic lectin from *Vicia sativa*. Eur J Biochem 113:319–325

Gebauer G, Schimpl A, Rüdiger H (1982) Lectin-binding proteins as potent mitogens for B-lymphocytes from nu/nu mice. Eur J Immunol 12:491–495

Genaud L, Guillot J, Bétail G, Coulet M (1982) Purification of lectins from *Ricinus communis* by combination of affinity and ion exchange chromatography and characterization of the isolated proteins. J Immunol Methods 49:323–332

Ghosh BN, Dasgupta B, Sircar PK (1981a) Bacto-agar – a binding matrix for purification of a lectin from *Butea monosperma*. Indian J Biochem Biophys 18:166–169 cited after Chem Abstr 96:20058q

Ghosh BN, Dasgupta B, Sircar PK (1981b) Purification of lectin from a tropical plant *Momordica dioica*. Indian J Exp Biol 19:253–255 cited after Chem Abstr 94:172626r

Gietl C, Ziegler H (1980) Distribution of carbohydrate-binding proteins in different tissues of *Robinia pseudoacacia*. Biochem Physiol Pflanz 175:58–66

60

Gietl C, Kauss H, Ziegler H (1979) Affinity chromatography of a lectin from *Robinia pseudoacacia* and demonstration of lectins in sieve-tube sap from other tree species. Planta 144:367–371

Gilboa-Garber N, Mizrahi L (1981) A new mitogenic D-galactose-philic lectin isolated from the seeds of the coral-tree *Erythrina corallodendron*. Comparison with *Glycine max* (soybean) and *Pseudomonas aeruginosa* lectins. Canad J Biochem 59:315–320

Goldstein IJ (1983) Lectins from *Griffonia simplicifolia* seeds. J Biosci 5 (Suppl 1) 65–71

Goldstein IJ, Hayes CE (1978) The lectins. Adv Carbohydr Chem Biochem 35:127–340

Goldstein IJ, Hughes RC, Monsigny M, Osawa T, Sharon N (1980) What should be called a lectin? Nature (London) 285:66

Gordon JA, Blumberg S, Lis H, Sharon N (1972) Purification of SBA by affinity chromatography. Methods Enzymol 288:365–368

Greenwood JS, Stinissen HM, Peumans WJ, Chrispeels MJ (1986) *Sambucus nigra* agglutinin is located in protein bodies in the phloem parenchyma of the bark. Planta 167:275–278

Grubhoffer L, Tichá M, Kocourek J (1981) Isolation and properties of a lectin from the seeds of hairy vetch *(Vicia villosa)*. Biochem J 195:623–626

Gupta BKD, Chatterjee-Ghose R, Sen A (1980) Purification and properties of mitogenic lectins from seeds of *Lathyrus sativus* (chickling vetch) Arch Biochem Biophys 201:137–146

Güran A, Tichá M, Filka K, Kocourek J (1983) Isolation and properties of a lectin from the seeds of Indian bean or lablab *(Dolichos lablab)*. Biochem J 209:653–657

Haaß D, Frey R, Thiesen M, Kauss H (1981) Partial purification of a hemagglutinin associated with cell walls from hypocotyls of *Vigna radiata*. Planta 151:490–496

Hankins CN, Shannon LM (1978) The physical and enzymatic properties of a phytohemagglutinin from mung beans. J Biol Chem 253:7791–8897

Hapner KD, Robbins JE (1979) Isolation and properties of a lectin from sainfoin *(Onobrychis viciifolia)*. Biochim Biophys Acta 580:186–197

Hemperly JJ, Hopp TP, Becker JW, Cunningham BA (1979) The chemical characterization of favin, a lectin isolated from *Vicia faba*. J Biol Chem 254:6803–6810

Hemperly JJ, Mostov KE, Cunningham BA (1982) In vitro translation and processing of a precursor form of favin, a lectin from *Vicia faba*. J Biol Chem 257:7903–7909

Higuchi M, Iwai K (1985) Purification and some properties of the basic lectin from winged bean seeds. Agric Biol Chem 49:391–398

Hopp TP, Hemperly JJ, Cunningham BA (1982) Amino acid sequence and variant forms of favin, a lectin from *Vicia faba*. J Biol Chem 257:4473–4483

Hořejší V (1979a) Properties of *Ulex europaeus* II lectin isolated by affinity chromatography. Biochim Biophys Acta 577:389–393

Hořejší V (1979b) Separation of *Ricinus communis* lectins by affinity chromatography. J Chromatogr 169:457–458

Hořejší V, Kocourek J (1973) Studies on phytohemagglutinins. XII. 0-Glycosyl polyacrylamide gels for affinity chromatography of phytohemagglutinins. Biochim Biophys Acta 297:346–351

Hořejší V, Kocourek J (1978a) Studies on lectin. XXXVII. Isolation and characterization of the lectin from jimsonweed seeds *(Datura stramonium)*. Biochim Biophys Acta 532:92–97

Hořejší V, Kocourek J (1978b) Studies on lectins. XXXVI. Properties of some lectins prepared by affinity chromatography on 0-glycosyl polyacrylamide gels. Biochim Biophys Acta 538:299–315

Hořejší V, Chaloupecká O, Kocourek J (1978a) Studies on lectins. XLIII. Isolation and characterization of the lectin from restharrow roots *(Ononis hircina)*. Biochim Biophys Acta 539:287–293

Hořejší V, Haškovec C, Kocourek J (1978b) Studies on lectins. XXXVIII. Isolation and characterization of the lectin from black locust bark *(Robinia pseudoacacia)*. Biochim Biophys Acta 532:98–104

Hořejší V, Tichá M, Novotný J, Kocourek J (1980) Studies on lectins. XLVII. Some properties of D-galactose-binding lectins isolated from the seeds of *Butea frondosa*, *Erythrina indica* and *Momordica charantia*. Biochim Biophys Acta 623:439–448

Horisberger M (1977) Polysaccharides immobilized in polyacrylamide gel as high-capacity adsorbents for *Bandeiraea simplicifolia* lectins and concanavalin A. Carbohydr Res 53:231–237

Iglesias JL, Lis H, Sharon N (1982) Purification and properties of a D-galactose/N-acetyl-D-galactosamine specific lectin from *Erythrina cristagalli*. Eur J Biochem 123:247–252

Indravathamma P, Seshadri HS (1980) Lectin from rice. J Biosci 2:29–36 cited after Chem Abstr 96:120644s

Ito Y (1986) Occurrence of heparin inhibitable hemagglutinating activity in leaves of fava bean *(Vicia faba)*. Plant Sci 45:27–29

Iyer PNS, Wilkinson KD, Goldstein IJ (1976) An N-acetyl-D-glucosamine-binding lectin from *Bandeiraea simplicifolia* seeds. Arch Biochem Biophys 177:330–333

Janssen E, Uhlenbruck G (1983) Two "incomplete" anti-A lectins and "anti-N" blood group specificity in *Molucella laevis* seeds. Ärztl Lab 29:255–258

Joubert FJ, Sharon N, Merrifield EH (1986) Purification and properties of a lectin from *Lonchocarpus capassa* (apple-leaf) seed. Phytochemistry 25:323–327

Kaladas PM, Poretz RD (1979) Purification and properties of a mitogenic lectin from *Wisteria floribunda* seeds. Biochemistry 18:4806–4812

Kamberger W (1978) Binding specificity and purification of *Medicago sativa* lectin. In: Hoffmann-Ostenhof D (ed) Affinity Chromatography. Pergamon Press, Oxford, pp 295–298

Karhi KK, Gahmberg CG (1980) Isolation and characterization of the blood group A-specific lectin from *Vicia cracca*. Biochim Biophys Acta 622:337–343

Kilpatrick DC (1980) Isolation of a lectin from the pericarp of potato *(Solanum tuberosum)* fruits. Biochem J 191:273–275

Kilpatrick DC (1982) Purification and some properties of a lectin from the fruit juice of the tomato *(Lycopersicon esculentum)*. Biochem J 185:269–272

Kilpatrick DC, Yeoman MM (1978) Purification of the lectin from *Datura stramonium*. Biochem J 175:1151–1153

Kilpatrick DC, Weston J, Urbaniak SJ (1983) Purification and separation of tomato isolectins by chromatofocusing. Anal Biochem 134:205–209

Kimura AK, Wigzell H, Holmquist G, Ersson B, Carlsson P (1979) Selective affinity fractionation of murine Cytotoxic T lymphocytes (CTL). Unique lectin specific binding of the CTL associated surface glycoprotein, T145. J Exp Med 149:473–484

Kochibe N (1986) Mannose binding lectins of *Vicia tetrasperma* seed and their immunological relationship to other legume lectins of similar specificity. Plant Cell Physiol 27:661–669

Kochibe N, Furukawa K (1980) Purification and properties of a novel fucose-specific hemagglutinin of *Aleuria aurantia*. Biochemistry 19:2841–2846

Kochibe N, Furukawa K (1982) *Aleuria aurantia* hemagglutinin. Methods Enzymol 83:373–377

Kocourek J, Hořejší V (1981) Defining a lectin. Nature (London) 290:188

Kocourek J, Jamieson GA, Votruba T, Hořejší V (1977) Studies on phytohemagglutinins. I. Some properties of the lectin of horse gram seeds *(Dolichos biflorus)*. Biochim Biophys Acta 500:344–360

Kohn J, Wilchek M (1982) A new approach (cyano transfer) for cyanogen bromide activation of Sepharose at neutral pH, which yields activated resins, free of interfering nitrogen derivatives. Biochem Biophys Res Commun 107:878–884

Kolberg J, Sletten K (1982) Purification and properties of a mitogenic lectin from *Lathyrus sativus* seeds. Biochim Biophys Acta 704:26–30

Kolberg J, Sollid L (1985) Lectin activity of gluten identified as wheat germ agglutinin. Biochem Biophys Res Comm 130:867–872

Kolberg J, Michaelsen TE, Sletten K (1983) Properties of a lectin purified from the seeds of *Cicer arietinum*. Hoppe-Seyler's Z Physiol Chem 364:655–664

Konami Y, Tsuji T, Matsumoto I, Osawa T (1981) Purification and characterization of a *Cytisus*-type *Ulex europaeus* hemagglutinin II by affinity chromatography. Hoppe-Seyler's Z Physiol Chem 362:983–989

Konami Y, Yamamoto K, Tsuji T, Matsumoto I, Osawa T (1983a) Purification and characterization of two types of *Cytisus multiflorus* hemagglutinin by affinity chromatography. J Pharmacobio-Dyn 6:737–747 cited after Chem Abstr 100:20479z

Konami Y, Yamamoto K, Tsuji T, Matsumoto I, Osawa T (1983b) Purification and characterization of two types of Laburnum alpinum anti-H(0) hemagglutinin by affinity chromatography. Hoppe-Seyler's Z physiol Chem 364:397–405

Kortt AA (1984) Purification and properties of the basic lectins from winged bean seeds *(Psophocarpus tetragonolobus).* Eur J Biochem 138:519–525

Kortt AA (1985a) Characterization of the acidic lectins from winged bean seeds *(Psophocarpus tetra-gonolobus).* Arch Biochem Biophys 236:544–554

Kortt AA (1985b) Isolation of the acidic and basic lectins from winged bean seed *(Psophocarpus tetra-gonolobus).* J Sci Food Agric 36:863–870

Kortt AA (1986) Characterization of a lectin from the seeds of *Erythrina vespertilio.* Phytochemistry 25:2371–2374

Köttgen E, Volk B, Kluge F, Gerok W (1982) Gluten, a lectin with oligomannosyl specificity and the causative agent of gluten-sensitive enteropathy. Biochem Biophys Res Comm 109:168–173

Kouchalakos RN, Bates OJ, Bradshaw RA, Hapner KD (1984) Lectin from sainfoin *(Onobrychis viciifolia).* Complete amino acid sequence. Biochemistry 23:1824–1830

Kouchalakos RN, Hapner KD (1984) Carbohydrate specificity, metal content and molecular stability of a lectin from sainfoin *(Onobrychis viciifolia).* Biochim Biophys Acta 787:237–243

Kubánek J, Entlicher G, Kocourek J (1982) Studies on lectins. III. Isolation and characterization of the lectin from rye germ *(Secale cereale).* Acta biol med germ 41:771–780

Kumar GS, Appukuttan FS, Basu D (1982) α-D-Galactose-specific lectin from jack fruit *(Artocarpus integra)* seed. J Biosci 4:257–261 cited after Chem Abstr 98:2782d

Laemmli UK (1970) Cleavage of structural proteins during the assembly of the head of bacteriophage T4. Nature (London) 227:680–685

Lamb JE, Shibata S, Goldstein IJ (1983) Purification and characterization of *Griffonia simplicifolia* leaf lectins. Plant Physiol 71:879–887

Law IJ, Strijdom BW (1984) Properties of lectins in the root and seed of *Lotononis bainesii.* Plant Physiol 74:773–778

Lee S, Wang K (1983) Carbohydrate-binding specificity of a lectin from *Trichosanthes kirilowii.* Shengwu Huaxue Yu Shengwu Wuli Xuebao 15:139–142 cited after Chem Abstr 99:68709n

Lee S, Wang K, Wu K (1983) Porcine thyroglobulin-p-aminobenzylsulfonylethyl-cross-linked agar. An affinity adsorbent to purify a variety of lectins of different specificities. Shengwu Huaxue Yu Shengwu Wuli Yuebao 15:271–277 cited after Chem Abstr 99:174031v

Li SS-L (1980) Purification and partial characterization of two lectins from *Momordica charantia.* Experientia 36:524–527

Lin TT-S, Li SS-L (1980) Purification and physiochemical properties of ricins and agglutinins from *Ricinus communis.* Eur J Biochem 105:453–459

Lin J-Y, Lee T-C, Hu S-T, Tung T-C (1981) Isolation of four isotoxic proteins and one agglutinin from jequirity bean *(Abrus precatorius).* Toxicon 19:41–51

Lineweaver H, Murray CW (1947) Identification of the trypsin inhibitor of egg white with ovomucoid. J Biol Chem 171:565–581

Lis H, Sharon N (1972) Soybean *(Glycine max)* agglutinin. Methods Enzymol 288:360–365

Lis H, Sharon N (1981) Affinity chromatography for the purification of lectins (a review). J Chromatogr 215:361–372

Lis H, Joubert FJ, Sharon N (1985) Isolation and properties of N-acetyllactosamine-specific lectins from nine *Erythrina* species. Phytochemistry 24:2803–2809

Lisowska E, Szeliga W, Duk M (1976) Purification of *Vicia graminea* anti-N lectin by affinity chromatography. FEBS Letters 72:327–330

Lönngren J, Goldstein IJ, Bywater R (1976) Cross-linked guaran: a versatile immunosorbent for D-galactopyranosyl-binding lectins. FEBS Letters 68:31–34

Lotan R, Sharon N (1978) Peanut *(Arachis hypogaea)* agglutinin. Methods Enzymol 50:361–367

Lotan R, Skutelsky E, Danon D, Sharon N (1975) The purification, composition and specificity of the anti-T lectin from peanut *(Arachis hypogaea).* J Biol Chem 250:8518–8523

Luther P, Theise H, Chatterjee B, Karduck D, Uhlenbruck G (1980) The lectin from *Viscum album* – Isolation, characterization, properties and structure. Int J Biochem 11:429–435

Lutsik MD (1984) A new affinity sorbent for purification of lectins and its use for the isolation of wheat germ agglutinin. Ukr Biokim Zh 56:432–436

Lutsik MD, Antonyuk VA (1982) A new fucose-specific lectin from the bark of *Laburnum anagyroides*: Purification, properties and immunochemical specificity. Biokhimiya 47:1710–1715

Lutsik MD, Potapov MJ, Kirichenko NV (1977) Phytohemagglutinin anti-N from the leaves of *Vicia uni- juga*. Probl Gematol Pereliv Krovi 22:48–52 cited after Chem Abstr 87:148734j

Lynn KR, Clevette-Radford NA (1986) Lectins from latices of *Euphorbia* and *Elaeophorbia* species. Phyto- chemistry 25:1553–1557

Majumdar T, Surolia A, (1979) A general method for isolation of galactopyranosyl-specific lectins. Indian J Biochem Biophys 16:200–203 cited after Chem Abstr 91:188988j

Malý P, Tichá M, Kocourek J (1985) Studies on lectins. LVIII. Sugar-binding properties as determined by affinity electrophoresis, of α-galactosidases from *Vicia faba* seeds possessing erythroagglutinating activity. J Chromatogr 347:343–350

Matsumoto I, Kitagaki H, Akai Y, Ito Y, Seno N (1981) Derivatization of epoxy-activated agarose with vari- ous carbohydrates for the preparation of stable and high-capacity affinity adsorbents: their use for affinity chromatography of carbohydrate-binding proteins. Anal Biochem 116:103–110

Matsumoto I, Jimbo A, Mizuno Y, Seno N, Jeanloz RW (1983) Purification and characterization of potato lectin. J Biol chem 258:2886–2891

Maurer HR (1971) Disc electrophoresis and related techniques of polyacrylamide gel electrophoresis. de Gruyter Berlin (W)

Maylié-Pfenninger MF, Jamieson JD (1979) Distribution of cell surface saccharides on pancreatic cells. I. General methods for preparation and purification of lectins and lectin-ferritin conjugates. J Cell Biol 80:69–76

Mbadiwe EI, Agogbua SIO (1978) An anti-B specific hemagglutinin from the seeds of *Mucuna flagellipas*. Phytochemistry 17:1057–1058

McCurrach PM, Kilpatrick DC (1986) Purification of potato lectin (*Solanum tuberosum* agglutinin) from tubers or fruits using chromatofocusing. Anal Biochem 154:492–496

McPherson A, Hoover S (1979) Purification of mitogenic proteins from *Hura crepitans* and *Robinia pseudoacacia*. Biochem Biophys Res Comm 89:713–720

Meade NA, Staat RH, Langley SD, Doyle RJ (1980) Lectin-like activity from *Persea americana*. Carbohydr Res 78:349–363

Miller RL (1983) Purification of peanut (*Arachis hypogaea*) agglutinin isolectins by chromatofocusing. Anal Biochem 131:438–446

Miller RC, Bowles DJ (1983) Interrelationships between Gramineae lectins. Planta 157:138–142

Moreira R de A, Ainouz IL (1981) Lectins from seeds of jack fruit (*Artocarpus integrifolia* L.): isolation and purification of two isolectins from the albumin fraction. Biol Plant 23:186–192

Moreira RA, Cavada BS (1984) Lectin from *Canavalia brasiliensis*; isolation, characterization and behavior during germination. Biol Plant 26:113–120

Moreira R de A, de Oliveira JTA (1983) Lectins from the genus *Artocarpus*. Biol Plant 25:343–348

Moreira RA, Barros ACH, Stewart JC, Pusztai A (1983) Isolation and characterization of a lectin from the seeds of *Dioclea grandiflora*. Planta 158:63–69

Murphy LA, Goldstein IJ (1978) *Bandeiraea simplicifolia* I isolectins. Methods Enzymol 50:345–349

Nachbar MS, Oppenheim JD (1982) Tomato (*L. esculentum*) lectin. Methods Enzymol 83:363–368

Nachbar MS, Oppenheim JD, Thomas JO (1980) Lectins in the U.S. diet. Isolation and characterization of a lectin from the tomato (*Lycopersicon esculentum*). J Biol Chem 255:2056–2061

Nakajima T, Furukawa K (1979) Anti-A agglutinin from *Falcata japonica*. Medicine and Biology 99:281 to 284

Namjuntra P, Muanwongyathi P, Chulavatuatol M (1985) A sperm-agglutinating lectin from seeds of Jack fruit (*Artocarpus heterophyllus*). Biochem Biophys Res Comm 128:833–839

Newman RA (1977) Heterogeneity among the anti-$T_F$ lectins derived from *Arachis hypogaea*. Hoppe- Seyler's Z Physiol Chem 358:1517–1520

Nsimba-Lubaki M, Peumans WJ (1986) Seasonal fluctuations of lectins in barks of elderberry (*Sambucus nigra*) and black locust (*Robinia pseudoacacia*). Plant Physiol 80:747–751

Nsimba-Lubaki M, Peumans WJ, Allen AK (1986) Isolation and characterization of glycoprotein lectins from the bark of three species of elder, *Sambucus ebulus*, *S. nigra* and *S. racemosa*. Planta 168:113–118

Obata F, Sakai R, Shiokawa H (1978) Preparation of concanavalin A consisting solely of intact subunits. J Biochem (Tokyo) 84:103–109

Ochoa JL, Kristiansen T (1978) Stroma: as an affinity adsorbent for non-inhibitable lectins. FEBS Letters 90:145–148

Ochoa JL, Kristiansen T (1982) Purification and partial characterization of an agglutinin from *Phaseolus coccineus* var. "Alubia". Biochim Biophys Acta 705:396–404

Oda Y, Minami K (1986) Isolation and characterization of a lectin from tulip bulbs, *Tulipa gesneriana*. Eur J Biochem 159:239–245

Ogata S-I, Kamachi Y, Arita Y, Sato M, Muramatsu T (1985) A re-examination of the isolectin components of the fucose-binding proteins of *Lotus tetragonolobus*. Carbohydr Res 144:297–304

Ohtani K, Shibata S, Misaki A (1980) Purification and characterization of tora-bean *(Phaseolus vulgaris)* lectin. J Biochem 87:407–416

Olsnes S (1978a) Ricin and *Ricinus* agglutinin, toxic lectins from castor beans. Methods Enzymol 50:330 to 335

Olsnes S (1978b) Toxic and nontoxic lectins from *Abrus precatorius*. Methods Enzymol 50:323–330

Olsnes S, Pihl A (1982) Toxic lectins and related proteins. In: Cohen, van Heyningen (eds) Molecular Action of Toxins and Viruses. Chapter 3. Elsevier Biomedical Press, Amsterdam, pp 51–105

Olsnes S, Haylett T, Refsnes K (1978) Purification and characterization of the highly toxic lectin modeccin. J Biol Chem 253:5069–5073

Olsnes S, Haylett T, Sandvig K (1982a) The toxic lectin modeccin. Methods Enzymol 83:357–362

Olsnes S, Stirpe F, Sandvig K, Pihl A (1982b) Isolation and characterization of viscumin, a toxic lectin from *Viscum album* (mistletoe) J Biol Chem 257:13263–13270

Olson MOJ, Liener IE (1967) Some physical and chemical properties of concanavalin A, the phytohemagglutinin of the Jack bean. Biochemistry 6:105–111

Osawa T, Trimura T, Kawaguchi T (1978) *Bauhinia purpurea* agglutinin. Methods Enzymol 50:367–372

Osborn TC, Ausloos KA, Brown JWS, Bliss FA (1983) Bean lectins. III. Evidence for greater complexity in the structural model of *Phaseolus vulgaris* lectin. Plant Sci Letters 31:193–203

Owens RJ, Northcote DH (1980) The purification of potato lectin by affinity chromatography on a fetuin-Sepharose matrix. Phytochemistry 19:1861–1862

Park KB, Lee KS, Kim JR, Kim YJ (1981) Purification and characterization of a lectin obtained from banha *(Pinellia ternata)* roots. Hanguk Saenghwa Hakhoe Chi 14:137–147 cited after Chem Abstr 95:110522r

Pere M, Pere D, Rougé P (1981) Isolement et étude des propriétés physicochimiques et biologiques des lectines d'*Hura crepitans*. Planta Medica 41:344–350

Pereira MEA, Gruezo F, Kabat EA (1979) Purification and characterization of lectin II from *Ulex europaeus* seeds and an immunochemical study of its combining site. Arch Biochem Biophys 194:511–525

Perez G (1984) Isolation and characterization of a lectin from the seeds of *Erythrina edulis*. Phytochemistry 23:1229–1232

Petryniak J, Janusz M, Markowska E, Lisowska E (1981) Purification of the *Euonymus europaeus* lectin by affinity chromatography on the desialized MN blood group glycoprotein, and lectin $NH_2$-terminal analysis. Acta biochem polon 28:267–273

Peumans WJ, Stinissen HM (1982) Rice lectin: purification, properties, molecular structure and relationship to cereal lectins. Arch Intern Physiol Biochem 90:B210–B211

Peumans WJ, Spaepen C, Stinissen HM, Carlier AR (1982a) Isolation and partial characterization of a lectin from a false brome grass *(Brachypodium sylvaticum)*. Biochem J 205:635–638

Peumans WJ, Stinissen HM, Carlier AR (1982b) Isolation and partial characterization of wheat germ agglutinin-like lectins from rye *(Secale cereale)* and barley (Hordeum vulgare) embryos. Biochem J 203:239–243

Peumans WJ, Stinissen HM, Carlier AR (1982c) Subunit exchange between lectins from different cereal species. Planta 154:568–572

Peumans WJ, de Ley M, Broekaert WF (1984a) An unusual lectin from stinging nettle *(Urtica dioica)* rhizomes. FEBS Letters 177:99–103

65

Peumans WJ, Nsimba-Lubaki M, Carlier AR, van Driessche E (1984b) A lectin from *Bryonia dioica* root stocks. Planta 160:222–228

Peumans WJ, de Ley M, Stinissen HM, Broekaert WE (1985a) Isolation and partial characterization of a new lectin from seeds of the greater celandine *(Chelidonium majus)*. Plant Physiol 78:379–383

Peumans WJ, Nsimba-Lubaki M, Peeters B, Broekaert WF (1985b) Isolation and partial characterization of a lectin from ground elder *(Aegopodium podagraria)* rhizomes. Planta 164:75–82

Poola I, Seshadri HS, Bhavanandan VP (1986) Purification and saccharide-binding characteristics of a rice lectin. Carbohydr Res 146:205–217

Porath J (1974) General methods and coupling procedures. Methods in Enzymol 34B:13–30

Prigent MJ, Lombart C, Asseraf A (1982) Affinity chromatography of *Vicia graminea* (blood group N) lectin using concanavalin A-Sepharose. FEBS Letters 150:85–88

Pueppke SG (1979) Purification and characterization of a lectin from seeds of the winged bean, *Psophocarpus tetragonolobus*. Biochim Biophys Acta 581:63–70

Pueppke SG, Benny UK, Hymowitz T (1982) Soybean lectin from seeds of the wild soybean, *Glycine soja*. Plant Science Letters 26:191–197

Pusztai A, Grant G, Stewart JC (1981) A new type of *Phaseolus vulgaris* (cv Pinto III) seed lectin: isolation and characterization. Biochim Biophys Acta 671:146–154

Quinn JM, Etzler ME (1984) Isolation and characterization of a new lectin-like protein from *Dolichos biflorus*. Plant Physiol 75 (1. Suppl):21

Ramstorp M, Mattiasson B (1982) Affinity chromatographic purification of lentil lectin using immobilized yeast cells. Appl Biochem Biotechnol 7:67–70

Read SM, Northcote DH (1983) Subunit structure and interactions of the phloem proteins of *Cucurbita maxima* (pumpkin). Eur J Biochem 134:561–569

Richardson M, Campos FDAP, Moreira RA, Ainouz IL, Begbie R, Watt WB, Pusztai A (1984) The complete amino acid sequence of the major α-subunit of the lectin from the seeds of *Dioclea grandiflora*. Eur J Biochem 144:101–111

Robertson BJ, Strength DR (1983) Characterization of a lectin from cowpeas. Prep Biochem 13:45–56

Roque-Barreira MC, Campos-Neto A (1985) Jacalin: an IgA-binding lectin. J Immunol 134:1740–1743

Rossi CA, Falasca A, Franceschi C, Stirpe F (1979) Carbohydrate-binding specificity and cell membrane interaction of *Vicia sativa* lectin. Bull Mol Biol Med 4:59–67

Rougé P, Chabert P (1983) Purification and properties of a lectin from *Lathyrus tingitanus* seeds. FEBS Letters 157:257–260

Rougé P, Sousa-Cavada B (1984) Isolation and partial characterization of two isolectins from *Lathyrus ochrus* seeds. Plant Sci Letters 37:21–27

Rüdiger H (1977) Purification and properties of blood group specific lectins from *Vicia cracca*. Eur J Biochem 72:317–322

Rüdiger H (1984) On the physiological role of plant lectins. Bioscience 34:95–99

Rutherford WM, Dick WE, Cavins JF, Dombrink-Kurtzman MA, Slodki ME (1986) Isolation and characterization of a soybean lectin having 4-0-methylglucuronic acid specificity. Biochemistry 25:952–958

Sabnis DD, Hart JW (1978) The isolation and some properties of a lectin (haemagglutinin) from *Cucurbita* phloem exudate. Planta 142:97–101

Schnaar RL, Lee YC (1975) Polyacrylamide gels copolymerized with active esters. A new medium for affinity systems. Biochemistry 14:1535–1541

Schurz H, Rüdiger H (1982) A spectrophotometric determination of protein immobilized to affinity gels. Anal Biochem 123:174–177

Shet MS, Murugiswamy B, Madaiah M (1985) A lectin from winged bean *(Psophocarpus tetragonolobus)* tubers. Indian J Biochem Biophys 22:313–315

Shibata S, Goldstein IJ, Baker DA (1982) Isolation and characterization of a Lewis b-active lectin from *Griffonia simplicifolia* seeds. J Biol Chem 257:9324–9329

Simmons BM, Russel JH (1985) A single affinity column step method for the purification of ricin toxin from castor beans (Ricinus communis). Anal Biochem 146:206–210

Solheim B (1983) Purification and characterization of lectins from *Vicia hirsuta* Physiol Plant 58:515–522

66

Sousa-Cavada B, Rougé P (1985) Partial characterization of two isolectins isolated from *Lathyrus cicera* seeds. Arq Biol Technol 28:421–430 cited after Chem Abstr 103:176836m

Stinissen HM, Peumans WJ, Carlier AR (1982) Posttranslational modification of rice lectin. Arch Intern Physiol Biochim 90:B216–B217

Strosberg AD, Lauwereys M, Foriers A (1983) Molecular evolution of legume lectins. In: Goldstein IJ, Etzler ME (eds) Chemical Taxonomy, Molecular Biology and Function of Plant Lectins. Alan R Liss, Inc, New York, pp 7–20

Sueyoshi S, Tsuji T, Osawa T (1985) Purification and characterization of four isolectins of mushroom (Agaricus bisporus). Biol Chem Hoppe-Seyler 366:213–221

Sumner JB, Howell SF (1936) The identification of the hemagglutinin of the Jack bean with concanavalin A. J Bacteriol 32:227–237

Sutoh K, Rosenfeld L, Lee YC (1977) Isolation of peanut lectin by affinity chromatography on polyacrylamide-entrapped guar beads and polyacrylamide (co-allyl-α-D-galactopyranoside). Anal Biochem 79:329–337

Tajima T, Tokita T, Sakurai Y, Gotoh A, Ikemoto S (1977) Agglutinin contained in a wild plant *Vicia unijuga* which shows an affinity against V- and N-antigens of goat and human red blood cells. Chikusan No Kenkyu 32:423–424 cited after Chem Abstr 87:100527s

Terao T, Irimura T, Osawa T (1975) Purification and characterization of a hemagglutinin from *Arachis hypogaea*. Hoppe-Seyler's Z physiol Chem 356:1685–1692

Tichá M, Zeineddine I, Kocourek J (1980) Studies on lectins. XLVIII. Isolation and characterization of lectins from the seeds of *Lathyrus odoratus* and *L. silvestris*. Acta biol med Germ 39:649–655

Tobiska J (1964) Die Phythämagglutinine. Akademie-Verlag, Berlin

Tollefsen SE, Kornfeld R (1983) Isolation and characterization of lectins from *Vicia villosa*. Two distinct carbohydrate-binding specificities are present in seed extracts. J Biol Chem 258:5165–5171

Tomita M, Kurokawa T, Onozaki K, Ichiki N, Osawa T, Ukita T (1972) Purification of galactose-binding phytoagglutinins and phytotoxins by affinity column chromatography using Sepharose. Experientia 28:84–85

Tovar J, Levy-Benshimol A, Seidl DS (1983) Crepitin isolectins in the latex of *Hura crepitans*. Acta Cient Venez 34:216–221 cited after Chem Abstr 102:218433q

Truchet GL, Sherwood JE, Pankratz HS, Dazzo FB (1986) Clover root exudate contains a particular form of the lectin, trifoliin A, which binds to *Rhizobium trifolii*. Physiol Plant 66:575–582

Tsuda M (1979) Purification and characterization of a lectin from rice bran. J Biochem 86:1451–1461

Turner RH, Liener IE (1975) The use of glutaraldehyde-treated erythrocytes for assaying the agglutinating activity of lectins. Anal Biochem 68:651–653

Uy R, Wold F (1977) 1.4-Butanediol diglycidylether coupling of carbohydrates to Sepharose: affinity adsorbents for lectins and glycosidases. Anal Biochem 81:98–107

Van Driessche E, Vandenbranden S, Kanarek L (1978) Improvement in the purification procedure of pea-lectin, and considerations on the subunit structure. Arch Intern Physiol Biochim 86:963–964

Vretblad P (1976) Purification of lectins by biospecific affinity chromatography. Biochim Biophys Acta 434:169–176

Wang K, Lee S, Wei Y (1983) Isolation and properties of lectin and trypsin inhibitor from *Trichosanthes kirilosii* by affinity chromatography. Shengwu Huaxue Yu Shengwu Wuli Xuebao 15:133–138 cited after Chem Abstr 99:66554j

Wantyghem J, Goulut C, Frénoy J-P, Goussault Y (1984) Lectins of *Robinia pseudoacacia*. In: Arnaud P (ed) Marker Proteins in Inflammation. de Gruyter, Berlin (W), pp 625–626

Wantyghem J, Goulut C, Frénoy J-P, Turpin E, Goussault Y (1986) Purification and characterization of *Robinia pseudoacacia* seed lectins. A re-investigation. Biochem J 237:483–489

Woods A, Hunter N, Sequeira L, Kelman A (1979) Lectin activity isolated from corn seed. Plant Physiol 63:8134

Yaakobovich Y, Neeman I (1983) Partial isolation and characterization of a hemagglutinating factor from avocado seeds. Arch Toxicol Suppl 6:52–57

Yamamoto S, Sakai I (1981) Composition and immunochemical properties of glycoproteins with anti-B agglutinin activity isolated from *Euonymus sieboldiana* seeds. J Immunogenet 8:271–279

67

Young NM, Jackson GED (1984) Anomalous behavior of lectins in size-exclusion high-performance liquid chromatography and gel electrophoresis. J Chromatogr 336:397–402

Young NM, Leon MA (1978) Preparation of affinity-chromatography media from soluble polysaccharides by cross-linkage with divinyl sulfone. Carbohydr Res 66:299–302

Young NM, Williams RE, Roy C, Yaguchi M (1982) Structural comparison of the lectin from sainfoin *(Onobrychis viciifolia)* with Con A and other D-mannose-specific lectins. Canad J Biochem 60:933–941

Young NM, Watson DC, Williams RE (1984) Structural differences between two lectins from *Cytisus scoparius*, both specific for D-galactose and N-acetyl-D-galactosamine. Biochem J 222:41–48

Young NM, Watson DC, Williams RE (1985) Lectins and legume chemotaxonomy: Characterization of the N-acetyl-D-galactosamine specific lectin of *Bauhinia purpurea*. FEBS Letters 182:403–406

Zenteno E, Ochoa JL (1985) *Cacti* lectins. In: Bøg-Hansen TC, Breborowicz J (eds) Lectins. de Gruyter, Berlin (W), Vol 4, pp 437–445

Zheng Z, Deng J, Xiac Z, Cai J, Zhang H (1985) Purification and partial characterization of the lectin from *Vicia bungei*. Shengwu Huaxue Zazhi 1:73–79 cited after Chem Abstr 104:86880n

Zhu Z, Sun C (1982) Purification and properties of sesame lectin. Shengwu Huaxue Yu Shengwu Wuli Xuebao 14:91–93 cited after Chem Abstr 97:90252p

Ziska P, Franz H (1982) The lectin from garden cress *(Lepidium sativum)*. Isolation and characterization. In: Bøg-Hansen TC, Spengler G (eds) Lectins – Biology, Biochemistry, Clinical Biochemistry. de Gruyter, Berlin (W), Vol 2, pp 711–719

# Additions in Print

In the meantime after writing the first text to that book the author added some remarks reflecting the latest results in this particular field of research. We'll add it behind the original work.

### Allium (Liliaceae)
Unspecific agglutinating activities were found in bulbs and leaves of several *Allium* species (Sun and Yu 1986). The lectin from *Allium sativum* which reacts with mannose was purified by an unspecified affinity adsorbent. The preparation consisted of two subunits of 26,500 and 47,500 MW.

### Amaranthus cruentus (Amaranthaceae)
Vazquez-Moreno and Calderon de la Barca (1987) detected a mitogenic GalNAc-binding lectin in the seeds. It was purified on fetuin which had been immobilized to a commercial vinylsulfone agarose. The subunit size of the lectin is 32,500 MW.

### to Amphicarpea bracteata
Recently, this lectin has been reinvestigated by Maliarik et al. (1987). The authors used a commercial adsorbent which carried a synthetic blood group A trisaccharide. The lectin was eluted by GalNAc. In gel filtration, the molecular weight of the lectin was determined to be 135,000 in gel electrophoresis, regardless whether or not in the presence of mercaptoethanol, four bands (32,000, 30,000, 28,000, 27,500 MW) appeared all of which contain carbohydrate.

### Arisaema wilsonii (Araceae)
From the tubers, a lectin was isolated by the use of immobilized thyroglobulin (Xiao and Zheng 1985). It is composed of two subunit types (24,000 and 11,000 MW). Though it agglutinates only rabbit but not human red blood cells, it acts as a mitogen towards human lymphocytes.

### to Artocarpus

More recently, the *A. integrifolia* lectin has also been purified on immobilized IgA (Roque-Barreira et al. 1986) and GalNAc (Vijayakumar and Forrester 1986). Purification protocols for other species as *A. altilis* (Hunter et al. 1986a; Hunter et al. 1986b), *A. lakoocha* (Chowdhury et al. 1987) and *A. tonkinensis* (Khang et al. 1987) include affinity chromatography on immobilized monosaccharides, oligosaccharides or glycoproteins or ion exchange and gel filtration. The preparations obtained from different species and in different laboratories vary in molecular weight (mostly about 43,000 to 45,000 MW and subunit compositions (between 11,000 and 18,000 MW).

### Clerodendron trichotomum (Verbenaceae)

From the fruit of this plant, Kitagaki et al. (1985) isolated a lectin by means of the p-aminophenyl glycosides of β-Gal and α-GalNAc immobilized to agarose. The glycoprotein lectin has a molecular weight of about 56,000 and consists of two equal subunits. Later, the same group improved the procedure by using lactamyl-Sepharose (Kitagaki-Ogawa et al. 1986). Though this adsorbent is lower in capacity it does not bind dyes which occur in the fruits and thus obviates a pretreatment of the extract. A similar lectin occurs in the leaves (Kitagaki-Ogawa et al. 1986).

### Colchicum autumnale (Liliaceae)

Meadow saffron; Herbstzeitlose; colchique

The tubers contain a lectin which was adsorbed to fetuin-agarose and eluted by unbuffered diaminopropane (Peumans et al. 1986). In gel filtration, the lectin eluted at 100,000 MW in electrophoresis, two subunit types (15,000 and 10,000 MW) were seen. This indicates an octameric structure. Both rabbit and human red blood cells react with the lectin. Agglutination by the latter ones can be inhibited by lactose, galactose and some related sugars which surprisingly are ineffective with rabbit cells.

### to Euphorbia

From the latices of three *E.* species, Nsimba-Lubaki et al. (1986) isolated GalNAc-binding lectins by means of immobilized fetuin-agarose. These lectins were found to have a molecular weight of about 140,000 and to be composed of four different subunits.

### Galactia tenuiflora (Leguminosae)

Le Pendu et al. (1986) isolated the seed lectin by the use of a commercial adsorbent which carries the blood group H type 2 trisaccharide. The lectin was eluted with 2.5 M NH₃. In electrophoresis under non-reducing conditions, a diffuse band between 50,000 and 55,000 MW appeared, after reduction, two bands (27,000 and 29,000 MW) were seen. The lectin reacts strongly with blood group H type 2 but only weakly with the types 3 and 4 and not at all with type 1. It thus differs from the blood group H specific lectin from *Ulex europaeus*.

### to Galanthus nivalis

In a more recent paper, van Damme et al. (1987) report a subunit size of 13,000 MW, a total molecular weight of about 50,000 and an exclusive specificity towards mannose.

### Listera ovata (Orchidaceae)

Twayblade; Zweiblatt

In a short paper, van Damme and Peumans (1986) report the isolation of two isolectins from the leaves by means of immobilized mannose. Only rabbit but not human red blood cells are agglutinated. The only monosaccharide acting as an inhibitor is mannose. The total molecular weight of both lectins is 25,000, subunit sizes are 11,000 and 12,000.

### Luffa acutangula (Cucurbitaceae)

Ridge gourd

The exudate contains a lectin which was purified by affinity chromatography on immobilized glycopeptide from the soybean lectin. The *Luffa* lectin is a non-covalent dimer from 24,000 MW subunits. It does not bind any of the monosaccharides tested but interacts with chitooligosaccharides (see Anathara et al. 1986).

### Marah macrocarpus (Cucurbitaceae)

The root stock lectin was isolated by Peumans et al. (1987) on immobilized fetuin essentially in the same manner that was used for the *Bryonia dioica* lectin (Peumans et al. 1984b, see above). The *Marah macrocarpus* lectin has a molecular weight of 64,000 and is composed of two slightly different subunits 34,000 and 31,000 MW. It is very closely related to the *Bryonia dioica* lectin.

### Phoradendron californicum (Viscaceae)

Starting from their experience with the *Viscum album* lectins, Franz et al. (1987) isolated a lectin from the related plant using affinity chromatography on partially hydrolyzed agarose or on lactosyl-Sepharose. The total molecular weight is 68,000, after reduction it dissociates into an A- (31,000 MW) and a B-subunit (38,000 MW) as do many other toxic lectins (e. g. from *Viscum album* and *Ricinus communis*). Though the *Phoradendron* lectin inhibits protein synthesis in cells it is much less toxic towards mice than the *Viscum album* or *Ricinus communis* lectins.

### Salvia sclarae (Lamiaceae)

The seeds were known to specifically agglutinate $T_n$-erythrocytes. Piller et al. (1986) isolated the lectin essentially by solvent precipitation and ion exchange. A final purification was achieved by affinity chromatography on a commercial resin carrying α-linked GalNAc. The lectin has a total molecular weight of 50,000, after reduction, only one band at 35,000 is seen. This discrepancy may be attributed to the high (15 %) sugar content. Of the red blood cells tested, only $T_n$-cells were agglutinated effectively, against other cells, the lectin showed only little or no activity. The specificity was determined in some details, of the monosaccharides tested, GalNAc was the inhibitor of agglutination.

### to Sophora japonica

Lectins with similar sugar specificities but different subunit sizes were found by Ito (1986) in leaves and flowers.

### Tetracarpidium conophorum (Euphorbiaceae)

The seeds contain a mitogenic lectin that binds to galactose and its derivatives. It was purified by affinity chromatography on immobilized lactose. The monomeric lectin consists of a single chain (34,000 MW) (see Togun et al. 1987).

**to Tulipa**

More recently Oda et al. (1987) isolated a second lectin from the bulbs of *T. gesneriana*. In this case, immobilized thyroglobulin was used as an affinity adsorbent. Both lectins differ in their molecular sizes. The second lectin reacted only with mouse and rat but not with rabbit, sheep and human erythrocytes and with yeast cells. Some glycoproteins but no monosaccharide inhibited the agglutination.

Independently, Cammue et al. (1986) isolated a lectin from the bulbs of a hybrid cultivar by means of affinity chromatography on immobilized fetuin. In its molecular properties, this lectin differed from both lectins described by the Japanese group. It was most active with trypsinized rabbit erythrocytes but showed also a slight activity with human cells. Several monosaccharides were inhibitory GalNAc being the most active.

# References

Anathara V, Patanjali SR, Swamy MJ, Sanadi AR, Goldstein IJ, Surolia A (1986) Isolation, macromolecular properties and combining site of a chitooligosaccharide-specific lectin from the exudate of rigde gourd *(Luffa acutangula)*. J Biol Chem 261:14621–14627

Cammue BPA, Peeters B, Peumans WJ (1986) A new lectin from tulip *(Tulipa)* bulbs. Planta 169:583–588

Chowdhury S, Hafiz A, Chatterjee BP (1987) Purification and characterization of an α-galactosyl-binding lectin from *Artocarpus lakoocha* seeds. Carbohydr Res 159:137–148

Franz H, Müller P, Kindt A, Ziska P (1987) Viscaceae lectins: a new lectin from *Phoradendron californicum*. Proc 9th International Lectin Conference, Cambrigde (England), p 47

Hunter JB, Suresh MR, Noujaim AA, Hagen DS, Heeley DH, Micetich RG (1986a) Isolation and characterization of a lectin from breadfruit *(Artocarpus altilis)* seeds. Biochem Arch 2:319–328 cited after Chem Abstr 106:63216c

Hunter JB, Suresh MR, Keshavarz E, Wenman WM, Micetich RG (1986b) Purification of lectins from *Artocarpus altilis* and *Ficus deltoides* by gel filtration fast protein liquid chromatography. Biochem Arch 2:99 to 105 cited after Chem Abstr 105:59065q

Ito Y (1986) Occurence of lectins in leaves and flowers of *Sophora japonica*. Plant Science 47:77–82

Khang NQ, Strosberg AD, Hoebeke J (1987) Purification and physico-chemical properties of a lectin from *Artocarpus tonkinensis*. Proc 9th International Lectin Conference, Cambrigde (England), p 51

Kitagaki H, Seno N, Yamaguchi H, Matsumoto I (1985) Isolation and characterization of a lectin from the fruit of *Clerodendron trichotomum*. J Biochem 97:791–799

Kitagaki-Ogawa H, Matsumoto I, Seno N, Takahashi N, Endo S, Arata Y (1986) Characterization of the carbohydrate moiety of *Clerodendron trichotomum* lectins. Its structure and reactivity towards plant lectins. Eur J Biochem 161:779–785

Le Pendu J, Gérard G, Lambert F, Mollicone R, Oriol R (1986) A new anti-H lectin from the seeds of *Galactia tenuiflora*. Glycoconjugate J 3:203–216

Maliarik MJ, Roberts DD, Goldstein IJ (1987) Properties of the lectin from the hog peanut *(Amphicarpaea bracteata)*. Arch Biochem Biophys 255:194–200

Nsimba-Lubaki M, Allen AK, Peumans WJ (1986) Isolation and partial characterization of latex lectins from three species of the genus *Euphorbia* (Euphorbiaceae). Physiol Plant 67:193–198

Oda Y, Minami K (1986) Isolation and characterization of a lectin from tulip bulbs, *Tulipa gesneriana*. Eur J Biochem 159:239–245

Oda Y, Minami K, Ichida S, Aonuma S (1987) A new agglutinin from the *Tulipa gesneriana* bulbs. Eur J Biochem 165:297–302

Peumans WJ, Allen AK, Cammue BPA (1986) A new lectin from meadow saffron *(colchicum autumnale)*. Plant Physiol 82:1036–1039

Peumans WJ, Allen AK, Nsimba-Lubaki M, Chrispeels MJ (1987) Related glycoprotein lectins from root stocks of wild cucumbers. Phytochemistry 26:909–912

Piller V, Piller F, Cartron J-P (1986) Isolation and characterization of an N-acetylgalactosamine specific lectin from *Salvia sclarea* seeds. J Biol Chem 261:14069–14075

Roque-Barreira MC, Praz F, Halbwachs-Mecarelli L, Greene LJ, Campos-Neto A (1986) IgA-affinity purification and characterization of the lectin jacalin. Braz J Med Biol Res 19:149–157 cited after Chem Abstr 105:151305r

Sun C, Yu L (1986) Lectins from *Allium* plants. Shengwu Huaxue Yu Shengwu Wuli Xuebao 18:213–215 cited after Chem Abstr 105:131940r

Togun RA, Animashaun T, Kay JE (1987) Purification and characterization of a mitogenic galactose-binding lectin from *Tetracarpidium conophorum* seeds. Proc 9th International Lectin Conference, Cambrigde (England), p 81

Van Damme EJM, Peumans WJ (1986) A mannose-specific lectin from the leaves of the orchid twayblade *(Listera ovata)*. Arch Intern Physiol Biochim 95:B101

Van Damme EJM, Allen AK, Peumans WJ (1987) Isolation and characterization of a lectin with exclusive specificity towards mannose from snowdrop *(Galanthus nivalis)* bulbs. FEBS Letters 215:140–144

Vazquez-Moreno L, Calderon de la Barca AM (1987) Comparison of different matrix materials to affinity purify *Amaranthus cruentus* lectin. Proc 9th International Lectin Conference, Cambrigde (England), p 61

Vijayakumar T, Forrester JA (1986) Purification and physico-chemical properties of the lectins from jack fruit *(Artocarpus integrifolia)*. Biol Plant 28:370–374 cited after Chem Abstr 106:45997d

Xiao Z, Zheng Z (1985) Purification and properties of lectin from *Arisaema wilsonii*. Sichuan Daxue Xuebao, Ziran Kexueban: 92–98 cited after Chem Abstr 104:223314x

# 3 Structure and Function of Leguminosae Lectins
## Edilbert van Driessche

## 3.1 Introduction

The first time he noticed that *Ricinus communis* seed extracts are able to agglutinate red blood cells, Stillmark (1888) could hardly have imagined that his discovery was the start of a new field of research, now generally referred to as "Lectinology". Although it has long been assumed that lectins are exclusively confined to higher plants, it is now well established that these proteins are widely distributed in nature. They have been demonstrated in, and isolated from, bacteria, algae, fungi, higher plants, invertebrates, as well as vertebrate tissues (Goldstein and Hayes 1978; Lis and Sharon 1981; Rüdiger 1984). Today, over 100 lectins have been purified and characterized, some of them only superficially, but others in great detail.

Especially the seeds of legume plants have proven to be excellent lectin sources, since up to 15 % of the total protein content of these seeds may consist of lectin. This might be a principal reason why legume lectins have by far been mostly thoroughly investigated.

The first part of this chapter deals with the structural properties, biosynthesis and processing of Leguminosae lectins, while in the second part different hypotheses with respect to their physiological role are considered.

## 3.2 The Structure of Legume Lectins

### 3.2.1 Structural Studies on Concanavalin A

Without any doubt, from all lectins isolated until now, that from the seeds of *Canavalina ensiformis* (i.e., concanavalin A, Con A) has been studied in most detail. Indeed, Con A was the first lectin from which the primary structure (Wang et al. 1975; Cunningham et al. 1975) as well as the three-dimensional structure (Becker et al. 1975; Reeke et al. 1975) has been unraveled. Con A is a tetrameric metalloprotein composed of four identical subunits, each containing 237 amino acids. Below pH 6, the predominant form of Con A is a dimer, while above pH 7 the tetramer predominates (Kalb and Lustig 1968). Each subunit contains one $Mn^{2+}$ ion and one $Ca^{2+}$ ion (Kalb and Levitzki 1968), has a single carbohydrate-binding site (Yariv et al. 1968; So and Goldstein 1968) and is devoid of any covalently bound carbohydrate, lipid or other detectable prosthetic groups (Olson and Liener 1967; Agrawal and Goldstein 1968). Besides the intact subunits, affinity purified Con A also contains fragmented subunits resulting from the cleavage of the peptide bond between residues 118–119 (Wang et al. 1975, 1971). Both residues are located on a loop that extends from the main body of the molecule and as such would be readily accessible to enzymatic cleavage (Wang et al. 1971). Since the structure near this natural cleavage point is stabilized by hydrogen bonding between adjacent strands of the rear β-structure (see below), the cleavage has no significant effect on the folding of the polypeptide chain (Edelman et al. 1972; Wang et al. 1975; Becker et al. 1975). This conclusion has been confirmed by X-ray studies on crystals grown from purified intact chains and crystals grown from the native mixture (Becker et al. 1975). However, in cleaved Con A, there may be a general loosening of the β-structure in the cleavage region which might be related to differences in solubility properties (Cunningham et al. 1972) and subunit aggregation of the two forms of Con A (McKenzie and Sawyer 1973).

The primary structure of Con A (Fig. 3.1) displays some remarkable features (Cunningham et al. 1975) in that the distribution of charged amino acid residues is more dense in the $NH_2$-terminal half of the polypeptide than in the COOH-terminal portion. The charged residues are clustered into three groups, i.e., a group of six negatively charged residues between $Asp^2$ and $Asp^{28}$, a group of positive charges between $Lys^{30}$ and $Arg^{60}$, and another cluster of negative charges between $Asp^{71}$ and $Glu^{87}$. Most of the charged residues are exposed on the surface of the monomer (Reeke et al. 1975), except for the first negatively charged cluster in which three residues ($Glu^8$, $Asp^{10}$ and $Asp^{19}$) are implicated in metal-ion binding (Edelman et al. 1972; Wang et al. 1975; Becker et al. 1975). On the other hand, the COOH-terminus contains a smaller number of charged residues which are generally counteracted by oppositely charged residues in their vicinity. While the tryptophanyl residues are evenly distributed along the polypeptide chain, six out of the seven tyrosines are found in the $NH_2$-terminal part, and all the phenylalanines are located between residue 111 and the COOH-terminus. The hydrophobic residues of the COOH-terminus form a cluster in the interior of the molecule (Reeke et al. 1975; Edelman et al. 1978).

The three-dimensional structure of Con A has been determined by X-ray diffraction at 2 Å-resolution (Becker et al. 1975; Reeke et al. 1975). The protomers are ellipsoidal domes of $42 \times 40 \times 39$ Å dimensions. The most striking feature of the structure of Con A is the presence of two large and entirely antiparallel β-structures which comprise more than half of the residues of the molecule (Reeke et al. 1975). The large amount of pleated sheet is consistent with earlier studies using circular dichroism and optical rotation dispersion (Pflumm et al. 1971; McCubbin et al. 1971; McKenzie et al. 1973). One of the pleated sheets, the "back" sheet, forms almost the entire back surface of the molecule as well as the rear of the o-iodophenyl-β-D-glucopyranoside-binding cavity and plays a major role in dimer as well as in tetramer formation (Edelman et al. 1972; Hardman and Ainsworth 1972; Becker et al. 1975). The entire sheet contains about 64 residues arranged in six antiparallel chains (residues 156−152 + 130−123, 103−106 + 108−116, 200−188, 48−56, 66−59, 73−79) and 14 residues in short connecting loops (Fig. 3.2).

Ellipsoid dimers, $84 \times 40 \times 39$ Å in size, are formed by noncovalent interactions between two protomers. The interaction between two subunits is isologous, i.e., the binding region is made up of two identical sets of atoms, one from each protomer. The residues participating in these contacts belong to four different regions of the structure: residues 87−90, 136−139 and 175−178 at the front of the molecule, and residues 117−132 including the final strand of the back β-structure of the monomers. Besides, the main chain amide nitrogen and carbonyl oxygen atoms of residues 125, 127 and 129 from the back β-structure form hydrogen bonds with complementary atoms of the back β-structure in the other member of the dimer (Reeke et al. 1975). As a result of these interactions, the back β-structures of the two members of the dimer are antiparallel and joined edge-to-edge with hydrogen bonds extending across the monomer-monomer interface. As such, the entire back of the dimer is built up of a single pleated sheet consisting of 12 antiparallel chains.

Like the dimer, the Con A tetramer is stabilized by interactions involving the back β-pleated sheets. In this case the predominant interactions arise from side chains projecting from the sheets of each protomer into the regions of dimer-dimer contact. Among the significant interactions are four pairs of salt links between Lys[114] and Lys[116] on one dimer and Glu[192] on the other dimer (Reeke et al. 1975). These interactions provide the basis for the dissociation of tetrameric Con A into carbohydrate-binding dimers upon derivatization of Con A with succinic anhydride or acetic anhydride (Gunther et al. 1973). Similarly, modification of the lysyl residues with maleic anhydride results in the formation of dimeric Con A as a result of the disruption of the salt bridges mentioned above (Young 1974). Besides, His[51] and His[121], both located in the dimer-dimer contact area, are linked to Ser[117] and Ser[208] respectively by hydrogen-bonded solvent bridges. Since the Con A tetramer undergoes dissociation in acidic solution (Kalb and Lustig 1968), it is most probable that the protonation of the imidazole moieties may alter the hydrogen-bonded networks, leading to dissociation of the tetramer into dimers.

A second β-structure, the "front" pleated sheet, contains about 55 residues arranged in seven antiparallel chains (residues 40–36, 23–30, 11–4, 208–215, 96–87, 169–176 + 179–181 and 143–139) (see Fig. 3.2). They divide the front half of the molecule into two regions which are devoid of regular secondary structure. The left hand region consisting of residues 138–168 is arranged in three loosely organized turns. The right hand region contains both the $NH_2$ and the COOH-terminal parts of the polypeptide chain at the front, the metal-binding region at the top, and the front wall of the o-iodophenyl-β-D-glucopyranoside binding cavity at the lower right of the molecule. Furthermore, this region also contains a small approximately helical structure between residues 81–84 (Becker et al. 1975).

The $Mn^{2+}$ ion and the $Ca^{2+}$ ion are bound at the top of the molecule and closely together (4.6 Å apart from each other), though at two separate sites, designated $S_1$ and $S_2$ (Becker et al. 1975). In the native protein $Mn^{2+}$ is complexed with the carboxyl oxygens of the side chains of Glu[8], Asp[10], Asp[19], to the nitrogen of the imidazole ring of His[24], and to two water molecules. One of these water molecules is involved in a hydrogen-bonding network extending to the carbonyl oxygen of Val[32] and the hydroxyl oxygen of Ser[34]. The other water molecule is situated at the inner end of a shallow depression which contains solvent and which extends to the surface of the molecule (Becker et al. 1975). The $Ca^{2+}$ ion on the other hand is bound to the side chains of Asp[10], Asn[14], Asp[19], Tyr[12], and to two water molecules. One water molecule is hydrogen-bonded to the carboxyl-group of Asp[208], and the other one to the carbonyl-group of Arg[228] (Becker et al. 1975).

In view of the predominance of carboxylic acids as complexing ligands for both $Mn^{2+}$ and $Ca^{2+}$, the removal of the metal ions of Con A at low pH (Kalb and Levitzki 1968) is most probably due to the protonation of these acidic groups. Abolition of the carbohydrate-binding capacity is observed by derivatization of the carboxyl-groups with glycine-methyl ester and carbodiimide (Hassing et al. 1971). In view of the fact that saccharide-binding is dependent on metal-binding (Kalb and Levitzki 1968), and that the protein-complexing residues in the metal-binding site are predominantly carboxyl-groups (Edelman et al. 1972; Wang et al. 1975; Becker et al. 1975), the loss of carbohydrate-binding upon derivatization with glycine-methyl ester could occur as the result of the abolition of the metal-binding capacity of Con A, rather than through direct action at the carbohydrate-binding site.

Based on X-ray data, Becker et al. (1975) suggested that demetallized Con A contains a precursor site for the transition metal, which might consist of the residues Glu[8], Asp[10] and His[24]. It is supposed that binding of $Mn^{2+}$ to the precursor site might induce a conformational change which brings Asp[19] into the correct position to interact with the $Mn^{2+}$ ion. Tyr[12], Asn[14] and Asp[19] would then become properly oriented with respect to residues 10, 208 and 228, thereby creating the $Ca^{2+}$ binding site. $Ca^{2+}$ might then stabilize the native conformation of the COOH-terminal

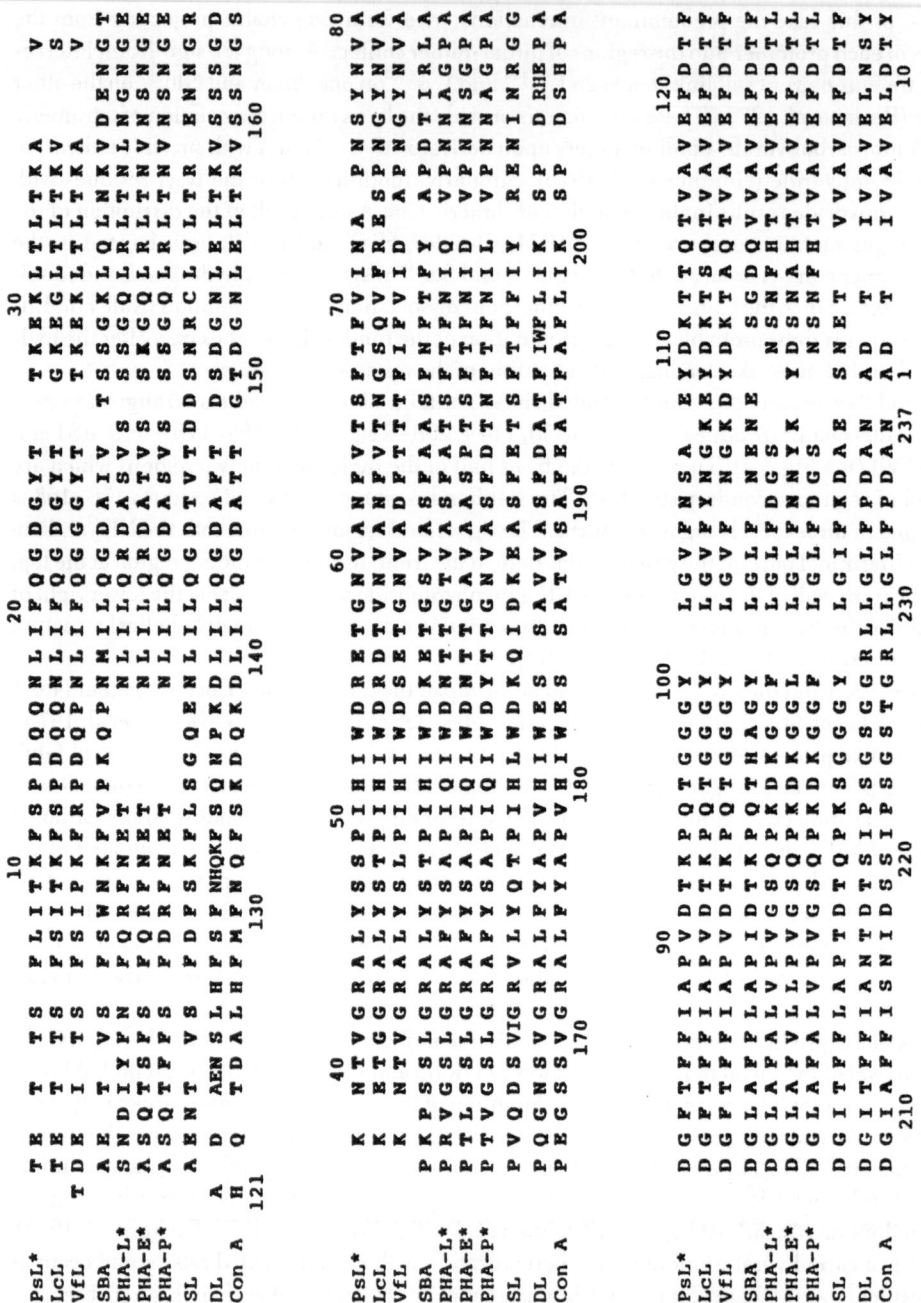

Fig. 3.1 Sequence homologies of PsL (Higgins et al. 1983), LcL (Foriers et al. 1978, 1981), VfL (Hopp et al. 1982), SBA (Vodkin et al. 1983), PHA-L and PHA-E (Hoffman and Donaldson 1985), PHA-P (Volker et al. 1986), SL (Kouchalakos et al. 1984), DL (Richardson et al. 1984a), Con A (Wang et al. 1975; Cunningham et al. 1975)

Numbers above the sequence correspond to amino acid residues in PsL (according to DNA sequence); numbers below correspond to amino acid residues in Con A. Deletions were introduced at appropriate positions in order to maximize the sequence homologies.

Abbreviations used:

PsL: Pisum sativum lectin    VfL: Vicia faba lectin    PHA-L: leukoagglutinin from Phaseolus vulgaris

LcL: Lens culinaris lectin    SBA: soybean agglutinin

PHA-E: erythroagglutinin from Phaseolus vulgaris    DL: Dioclea grandiflora lectin

PHA-P: phytohemagglutinin from Phaseolus vulgaris, var. Pinto III    Con A: concanavalin A (lectin from Canavalia ensiformis)

SL: sainfoin (Onobrychis viciifolia) lectin    *: sequence derived from cDNA

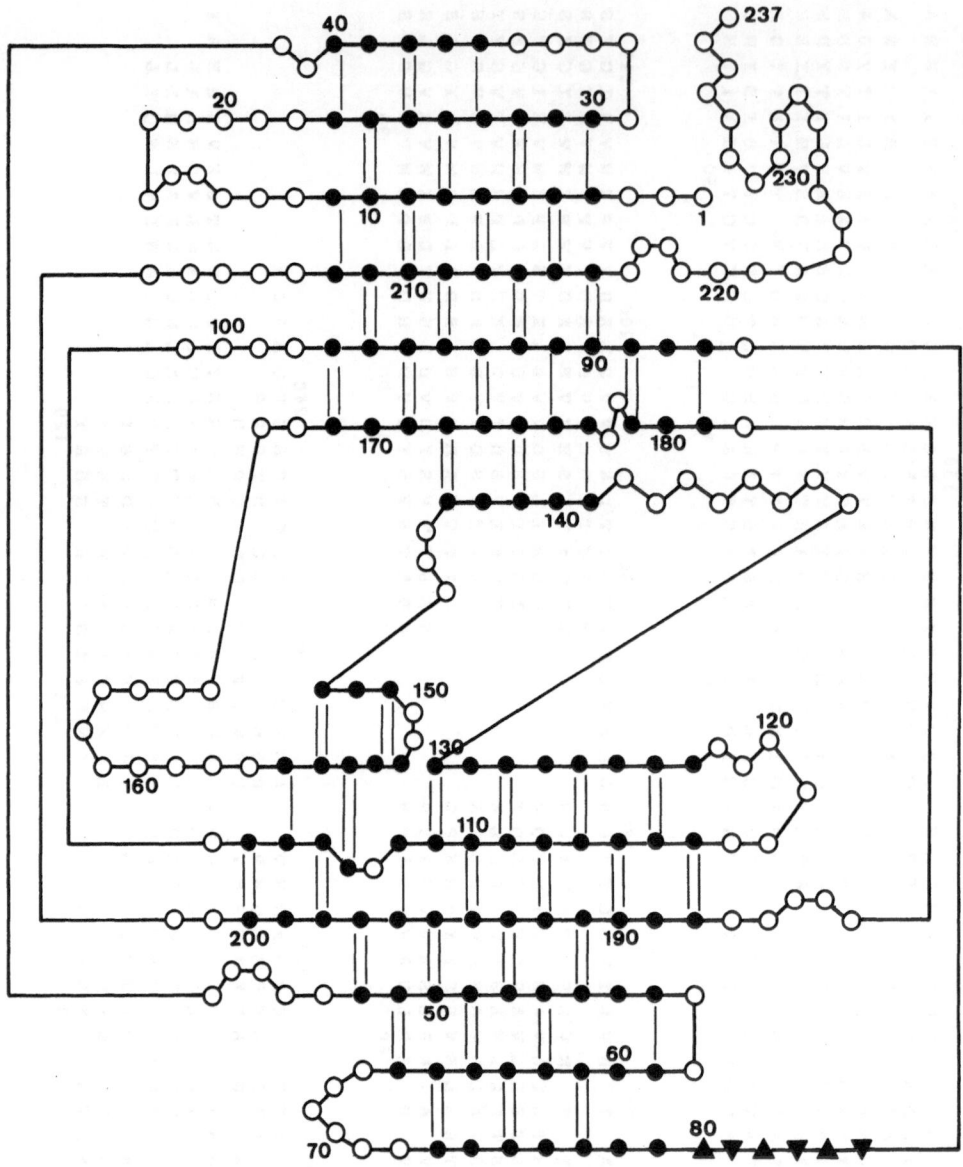

*Fig. 3.2* Schematic representation of the folding of the Con A polypeptide chain (simplified according to Reeke et al. 1975)

●, ▲: Residues implicated respectively in β-pleated sheet structures and in α-helix-like polypeptide folding

part of the chain, as well as the saccharide-binding site. On the other hand, Shoham et al. (1978) concluded from X-ray crystallographic studies that the large separation between the carboxyl-groups of $Asp^{19}$ and $Asp^{10}$ in demetallized Con A renders this pair of potential $Ca^{2+}$ complexing residues ineffective. Moreover, $Asp^{19}$ appears to form a salt bridge with $His^{24}$, and is therefore in an unfavorable position for ligand pairing with $Asp^{10}$. Binding of a transition metal-ion at $S_1$ deprotonates the $His^{24}$ residue (Gachelin et al. 1972) and consequently breaks the salt bridge with $Asp^{19}$, so that the carboxyl-group of the latter residue is free to assure a po-

78

sition relative to the carboxyl-group of $Asp^{10}$, which favors the formation of the $Ca^{2+}$ binding site. The studies mentioned above are in agreement with previous investigations indicating that $Mn^{2+}$ must be bound to demetallized Con A before the $Ca^{2+}$ ion, and that both sites must be occupied before saccharide-binding can occur (Kalb and Levitzki 1968; Yariv et al. 1968). Circular dichroism studies indicated that, upon sequential binding of metal-ions and saccharide, conformational changes occur affecting aromatic residues, although the gross secondary structure of the protein is apparently not affected (Pflumm et al. 1971; McCubbin et al. 1971).

Contrary to the unequivocal assignment of the metal-ion binding sites, the localization of the saccharide-binding site has long been a subject of controversy. Originally it was believed that the hydrophobic cavity of Con A represents the carbohydrate-binding site. This assumption was based on the observation (Becker et al. 1971) that o-iodophenyl-β-D-glucopyranoside binds in this hydrophobic cavity. Later it was concluded by Hardman and Ainsworth (1973) from X-ray crystallographic data, that a number of relatively nonpolar molecules such as methyl-p-hydroxybenzoate, o-iodobenzoic acid, o-iodoaniline, dimethyl-mercury and the noninhibiting carbohydrate derivative o-iodophenyl-β-D-galactopyranoside also bind into this site. Hardman and Ainsworth conclusively showed that saccharides labeled with heavy atoms such as o-iodophenyl-β-D-glucopyranoside bind to the hydrophobic cavity by their aglycones, rather than by their saccharide moieties. This conclusion was confirmed by Edelman and Wang (1978) who found that the binding of the nonpolar compounds β-indolacetic acid and tryptophan is independent of the saccharide-binding activity of Con A. A major problem during the localization of the saccharide-binding site was the dissolving or loss of the diffraction pattern of the Con A crystals when treated with high concentrations of inhibitory sugars (Becker et al. 1975). This problem could be overcome by Becker et al. (1976), who treated the crystals with glutaraldehyde. These authors could show that the residues 14–16, 97–99, 168–169, 207–208, 224–228 and 235–237 participate in Con A-saccharide interactions. These residues surround a shallow pocket near the top of the molecule in a region located at 10–15 Å from the metal-binding sites (Becker et al. 1976; Hardman and Ainsworth 1976).

As shortly mentioned above, each Con A protomer contains a large cavity extending deeply into the molecule, where several small hydrophobic molecules are known to bind (Hardman and Ainsworth 1973; Becker et al. 1975). The back of this cavity is made up of the large β-structure which extends across the rear of the molecule, while the front is delineated by the second β-structure, the short helical structure (residues 81–84) and a few sections of random coil. The cavity is built up of two distinct subsites. The first subsite is a large, predominantly hydrophobic region in which the iodine of o-iodophenyl-β-D-glucopyranoside binds. This region is surrounded by the side chains of $Tyr^{54}$, $Leu^{81}$, $Leu^{85}$, $Val^{89}$, $Val^{91}$, $Phe^{111}$, $Ser^{113}$, $Val^{179}$, $Ile^{181}$, $Phe^{191}$, $Phe^{212}$ and $Ile^{214}$. The second subsite is located between the first subsite and the surface of the molecule and contains predominantly hydrophylic groups such as the side chains of $Tyr^{54}$, $Ser^{56}$, $Asn^{82}$, $Ser^{113}$, $Ser^{189}$, as well as the main chain oxygen associated with $Lys^{114}$ and $Ile^{181}$ (Becker et al. 1975).

## 3.2.2  Structural Homologies between Legume Lectins

By now, the complete primary structure of several legume lectins has been determined (see Fig. 3.1), either by the classical Edman degradation procedure (Wang et al. 1975; Cunningham et al. 1975; Foriers et al. 1978, 1981; Cunningham et al. 1979; Hemperly et al. 1979; Hopp et al. 1982; Kouchalakos et al. 1984; Richardson et al. 1978, 1984a) or by DNA sequence analysis

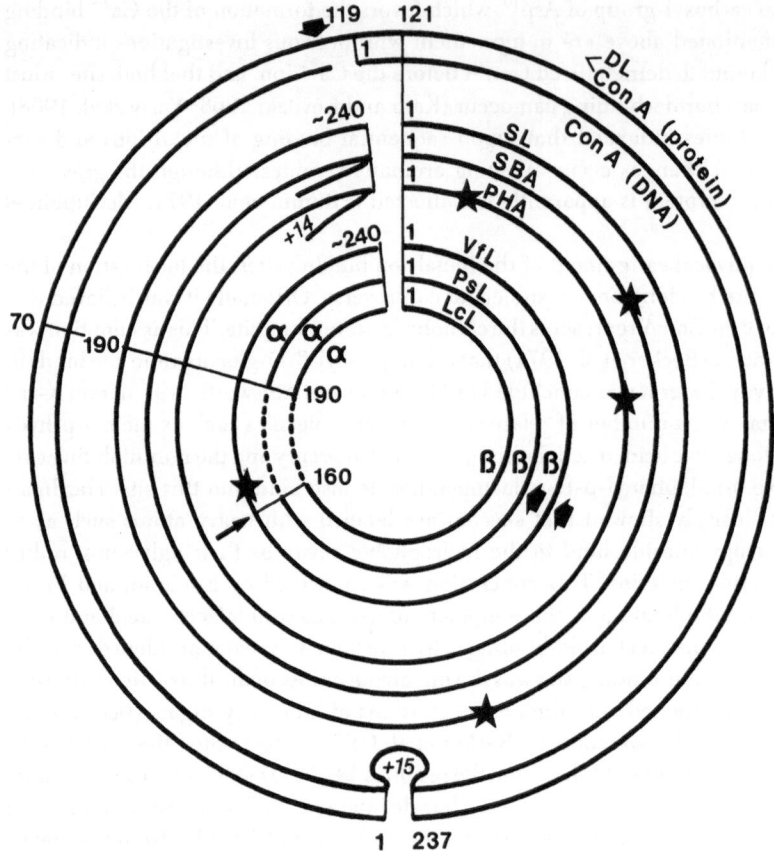

*Fig. 3.3* Schematic representation of two-chain lectins, one-chain lectins and Con A showing the circular permutation which gives maximal homology

α　: α-chain;　β: β-chain

－－－: stretches of missing amino acids in the protein sequence

✶　: putative glycosylation sites

➡　: natural cleavage sites

Abbrevations are as mentioned in Figure 1. For concanavalin A, the protein sequence ("Con A protein"), as well as the sequence derived from cDNA ("Con A DNA") are shown.

(Hoffman et al. 1982; Higgins et al. 1983a; Vodkin et al. 1983; Volker et al. 1986; Hoffman and Donaldson 1985). During the early stages of our own sequence work on pea-lectin, we noticed a striking homology of residues 13–29 of the α-chain of pea-lectin with sequence 80–99 in Con A (van Driessche et al. 1976a). Subsequently, Foriers et al. (1977a) compared the N-terminal sequences of both the α- and the β-chains of pea and lentil lectin: they showed that the α-chains of both lectins differ at only three positions out of the 25 which were determined, while the β-chains were identical except for two positions out of the 25 residues analyzed. Furthermore, sequence homology was observed between the N-terminal residues of the pea and lentil β-chains, and positions 123–147 of Con A. Moreover, Foriers et al. (1977b) found extensive sequence homology between the N-terminal amino acids of the β-chains of the two-chain lectins from pea and lentil on the one hand, and the amino terminal sequences of the one-chain lectins

80

from soybean and peanut and the E- and L-subunits of phytohemagglutinin on the other hand. From these results, a common genetic origin of the sequenced polypeptides was suggested and the hypothesis was put forward that the α- and β-chains of the two-chain lectins are derived, as the result of a proteolytic cleavage, from a precursor polypeptide chain with a molecular weight similar to that of the one-chain lectins (Fig. 3.3).

Upon unraveling the complete primary structure of favin, Hemperly et al. (1979) found that the sequence of the α-chain is homologous to a region in the middle of Con A (residues 70–119), while the β-chain was found to be homologous with two discrete segments of Con A: the homology starts at residue 120 of Con A, extends to the COOH-terminal (residue 237) and most remarkably continues without interruption through the 69 amino-terminal residues of Con A. From these observations Cunningham et al. (1979) concluded that favin and Con A are related to each other by a circular permutation of amino acid sequences (see Fig. 3.3). This circular permutation is also found for other legume lectins such as those isolated from *Lens culinaris*, *Phaseolus vulgaris* and *Glycine max* (see Fig. 3.3) and several genomic reorganizations, such as duplication and mutation, excision and reinsertion, have been proposed to be responsible for this feature (Cunningham et al. 1979; Foriers et al. 1981). Very recently it was found that the amino acid sequence derived from the cDNA of the Con A has a direct and not a circular homology with other legume lectins (Carrington et al. 1985), indicating that during the formation of mature Con A chains there has to be a transposition and a ligation of two peptides produced from a precursor polypeptide chain. Since the ligation to form mature Con A chains occurs between residues 118 and 119 (Carrington et al. 1985), it appears that the two fragments found in mature Con A are unligated polypeptides, rather than proteolytic degradation products of the polypeptide chain.

From the data in Table 3.1, it is obvious that all residues which are implicated in the binding of $Mn^{2+}$ and $Ca^{2+}$ are highly conserved in all the lectins mentioned. Similarly, the residues which line the hydrophobic cavity in Con A are fairly well conserved in the other lectins, or are mostly replaced by residues with similar properties. On the other hand, the amino acid residues which contribute to carbohydrate-binding in Con A are less well conserved in the other lectins. This is not surprising, since lectins such as Con A, LcL, PsL and Vfl, which are all classified as mannose/glucose specific lectins, and considered to be identical in terms of monosaccharide specificity, may nevertheless recognize different saccharide sequences (Debray et al. 1981; Cummings and Kornfeld 1982b).

Finally, the amino acid residues which form the back and the front β-structures of Con A are rather poorly conserved, especially those in the back pleated sheet (Table 3.2). However, strong β-formers (Met, Val, Ile) (Cou and Fassman 1974), β-formers (Cys, Tyr, Phe, Gln, Leu, Thr, Trp) or residues which are indifferent to β-forming (Arg, Gly, Asp) are mostly replaced by similar residues in the other lectins, suggesting that large parts of the polypeptide chain in these lectins too are folded in such a way as to form β-pleated sheets. This suggestion has been confirmed by circular dichroism studies. These studies have revealed that the lectin from the seeds of *Abrus precatorius* (Herrmann et al. 1978; Herrmann and Behneke 1980), *Arachis hypogaea* (Herrmann et al. 1978; Jirgensons 1978), *Bandeiraea simplicifolia* (Lönngren et al. 1976), *Cytisus scoparius* (Young et al. 1984), *Canavalia ensiformis* (Herrmann et al. 1978; McCubbin et al. 1971; Pflumm et al. 1971; Kay 1970; van Driessche, this paper), *Dolichos biflorus* (Père et al. 1975; Herrmann et al. 1978; Jirgensons 1980a, 1980b), *Glycine max* (Herrmann et al. 1978; Jirgensons 1978, 1980b), *Lens culinaris* (Herrmann et al. 1978; Jirgensons 1978; van Driessche, this paper), *Lotus tetragonolobus* (Herrmann et al. 1978; Jirgensons 1980a), *Onobrychis viciifolia* (Young et al. 1982), *Pisum sativum* (Herrmann et al. 1978; Jirgensons 1980a, 1980b; van Driessche et al. 1982), *Phaseolus vulgaris* (Herrmann et al. 1978; Jirgensons 1980a,

1980b), *Sophora japonica* (Jirgensons 1980a, 1980b), *Robinia pseudoacacia* (Père et al. 1975), *Vicia faba* (van Driessche, this paper), *Vicia sativa* (van Driessche, this paper), *Ulex europeus* (Herrmann et al. 1978; Jirgensons 1980a) all display circular dichroism spectra which in the far-UV region are strikingly similar in shape, wavelength of the minimum and in crossover point. From the magnitude of the ellipticities and the shape of the curves, it was concluded that the predominant ordered structure in those lectins is β-pleated sheet. Another common feature of most legume lectins is that saccharide-binding induces conformational changes affecting the microenvironment of the aromatic amino acid residues which contribute to the circular dichroism signal in the near-UV region (i.e., 250–320 nm) (Fig. 3.4, see pp. 87–93), but do not alter the secondary structure.

Further evidence for structural relatedness among legume lectins stems from immunological studies. By using Ouchterlony double immunodiffusion, Howard et al. (1979) showed that, with the exception of Con A and SBA, all legume lectins tested cross-react with heterologous legume-lectins antisera.

*Table 3.1* Sequence homologies between leguminous lectins: tabulation of the amino acid residues implicated in carbohydrate-binding and metal (Ca$^{2+}$ and Mn$^{2+}$)-binding and residues from the hydrophobic pocket.

Identity with the amino acid residue found in concanavalin A is marked with a plus (+) sign. Substitutions are as indicated; deletions are marked with a minus (−) sign. The amino acid sequences of PsL, PHA-L, PHA-E and PHA-P were derived from cDNA (see Fig. 3.1).

Boxes indicate sequence homologies either within or in between tribes.

| Con A | DL | PsL | LcL | VfL | SBA | PHA-L | PHA-E | PHA-P | SL |
|---|---|---|---|---|---|---|---|---|---|
| **Carbohydrate binding:** | | | | | | | | | |
| Asn– 14 | + | + | + | + | + | + | + | + | + |
| Thr– 15 | + | Ala | Ala | Ala | Ser | Lys | Val | Lys | Arg |
| Asp– 16 | + | Ala | Ala | Ala | − | + | His | + | − |
| Thr– 97 | + | + | + | + | + | + | + | + | + |
| Gly– 98 | + | + | + | + | + | + | + | + | + |
| Leu– 99 | + | Ala | Ala | Ala | + | Asn | Thr | Thr | Asp |
| Ser–168 | + | Asn | Glu | Asn | + | Gly | + | Gly | Asp |
| Ser–169 | + | Thr | Thr | Thr | + | + | + | + | + |
| Ala–207 | + | + | + | + | + | + | + | + | + |
| Asp–208 | + | + | + | + | + | + | + | + | + |
| Gly–224 | + | + | + | + | His | Lys | Lys | Lys | + |
| Ser–225 | + | Gly | Gly | Gly | Ala | Gly | Gly | Gly | Gly |
| Thr–226 | Gly | Gly | Gly | Gly | Gly | Gly | Gly | Gly | Gly |
| Gly–227 | + | Tyr | Tyr | Tyr | Tyr | Phe | Leu | Phe | Tyr |
| Arg–228 | + | − | − | − | − | − | − | − | − |
| Asp–235 | + | Ser | Asn | Asn | Asn | Gly | Asn | Gly | + |
| Ala–236 | + | + | Gly | Gly | Glu | Ser | Tyr | Ser | + |
| Asn–237 | + | − | Lys | Lys | + | − | Lys | − | Glu |

*Table 3.1* continuation

| Con A | DL | PsL | LcL | VfL | SBA | PHA-L | PHA-E | PHA-P | SL |
|---|---|---|---|---|---|---|---|---|---|
| **Metal binding: Ca$^{2+}$** | | | | | | | | | |
| Asp- 10 | Asn | + | + | + | + | + | + | + | + |
| Tyr- 12 | + | Phe | Phe | Phe | Phe | Leu | Leu | Leu | Phe |
| Asn- 14 | + | + | + | + | + | + | + | + | + |
| Asp- 19 | + | + | + | + | + | + | + | + | + |
| **Metal binding: Mn$^{2+}$** | | | | | | | | | |
| Glu- 8 | + | + | + | + | + | + | + | + | + |
| Asp- 10 | Asn | + | + | + | + | + | + | + | + |
| Asp- 19 | + | + | + | + | + | + | + | + | + |
| His- 24 | + | + | + | + | + | + | + | + | + |
| **Hydrophobic pocket:** | | | | | | | | | |
| Tyr- 54 | + | Phe | - | Phe | + | + | + | + | + |
| Leu- 81 | + | + | + | + | + | + | + | + | + |
| Leu- 85 | + | Val | Val | Val | + | + | + | + | + |
| Val- 89 | + | + | + | + | + | + | + | + | + |
| Val- 91 | + | Ile | Ile | Ile | Ile | + | + | + | Ile |
| Phe-111 | + | + | + | + | + | + | + | + | + |
| Ser-113 | + | + | + | + | + | + | + | + | + |
| Val-179 | + | Ile | Ile | Ile | Ile | Ile | Ile | Ile | Ile |
| Ile-181 | + | + | + | + | + | + | + | + | Leu |
| Phe-191 | + | + | + | + | + | + | Ser | + | + |
| Phe-212 | + | + | + | + | + | + | + | + | + |
| Ile-214 | + | + | + | + | Leu | Leu | Leu | Leu | Leu |
| Ser- 56 | + | Ala | - | Ala | Ala | + | + | + | Ala |
| Asn- 82 | + | Lys | Lys | Lys | Lys | Lys | Lys | Lys | Arg |

*Table 3.2* Sequence homologies between leguminous lectins: tabulation of the amino acid residues which are folded in concanavalin A as β-pleated sheet and α-helix-like segments (see also Fig. 3.3).

Symbols used are as in Table 3.1.

| Con A | DL | PsL | LcL | VfL | SBA | PHA-L | PHA-E | PHA-P | SL |
|---|---|---|---|---|---|---|---|---|---|
| **β-pleated sheets (residues of Con A front sheet):** | | | | | | | | | |
| Trp- 40 | + | + | + | + | + | + | + | + | + |
| Lys- 39 | Arg | Ser | Ser | Ser | Ser | Arg | Thr | Pro | Pro |
| Ala- 38 | + | Lys | Lys | Lys | Thr | Thr | Thr | Thr | Thr |
| Thr- 37 | + | + | + | + | + | + | + | + | + |
| Lys- 36 | Ser | Asn | Asn | Ser | + | + | + | + | Ile |

*Table 3.2*  continuation

| Con A | DL | PsL | LcL | VfL | SBA | PHA-L | PHA-E | PHA-P | SL |
|---|---|---|---|---|---|---|---|---|---|
| Pro- 23 | + | Arg | Arg | Arg | + | Arg | Arg | Arg | Ser |
| His- 24 | + | + | + | + | + | + | + | + | + |
| Ile- 25 | + | + | + | + | + | + | + | + | + |
| Gly- 26 | + | + | + | + | + | + | + | + | + |
| Ile- 27 | + | + | + | + | + | + | + | + | + |
| Asp- 28 | + | + | + | + | Asn | + | + | + | Asn |
| Ile- 29 | + | Val | Val | Val | Val | Val | Val | Val | Val |
| Lys- 30 | + | Asn | Asn | Asn | Asn | Asn | Asn | Asn | Asn |
| Thr- 11 | Ser | + | + | + | + | + | + | + | + |
| Asp- 10 | Asn | + | + | + | + | + | + | + | + |
| Leu-  9 | + | Phe | Phe | Phe | Phe | Phe | Phe | Phe | Phe |
| Glu-  8 | + | + | + | + | + | + | + | + | + |
| Val-  7 | + | + | + | + | + | + | + | + | + |
| Ala-  6 | + | + | + | + | + | + | + | + | + |
| Val-  5 | + | + | + | + | + | + | + | + | + |
| Ile-  4 | + | Thr | Thr | Thr | Val | Thr | Thr | Thr | Val |
| Asp-208 | + | + | + | + | + | + | + | + | + |
| Gly-209 | + | + | + | + | + | + | + | + | + |
| Ile-210 | + | Phe | Phe | Phe | Leu | Leu | Leu | Leu | + |
| Ala-211 | Thr | Thr | Thr | Thr | + | + | + | + | Thr |
| Phe-212 | + | + | + | + | + | + | + | + | + |
| Phe-213 | + | + | + | + | + | Ala | Val | Ala | + |
| Ile-214 | + | + | + | + | Leu | Leu | Leu | Leu | Leu |
| Ser-215 | Ala | Ala | Ala | Ala | Ala | Val | Leu | Val | Ala |
| Ser- 96 | Thr | Thr | Thr | Thr | Ala | Thr | Thr | Thr | Ala |
| Ala- 95 | + | + | + | + | + | + | + | + | + |
| Ser- 94 | + | + | + | + | + | + | Thr | + | + |
| Leu- 93 | + | Phe | Phe | Phe | Phe | Phe | Phe | Phe | + |
| Gly- 92 | + | + | + | + | + | + | + | + | + |
| Val- 91 | + | Ile | Ile | Ile | Ile | + | + | + | Ile |
| Arg- 90 | + | + | + | + | + | Ser | Ile | Ser | + |
| Val- 89 | + | + | + | + | + | + | + | + | + |
| Trp- 88 | + | + | + | + | + | + | + | + | + |
| Glu- 87 | + | + | + | + | + | + | + | + | Gln |
| Ser-169 | + | Thr | Thr | Thr | + | + | + | + | + |
| Val-170 | + | + | Gly | + | Leu | Leu | Leu | Leu | +/Ile |
| Gly-171 | + | + | + | + | + | + | + | + | + |
| Arg-172 | + | + | + | + | + | + | + | + | + |
| Ala-173 | + | + | + | + | + | + | + | + | Val |
| Leu-174 | + | + | + | + | + | Phe | Phe | Phe | + |
| Phe-175 | + | Tyr | Tyr | Tyr | Tyr | Tyr | Tyr | Tyr | Tyr |
| Tyr-176 | + | Ser | Ser | Ser | Ser | Ser | Ser | Ser | Gln |
| Val-179 | + | Ile | Ile | Ile | Ile | Ile | Ile | Ile | Ile |
| His-180 | + | + | + | + | + | Gln | Gln | Gln | + |
| Ile-181 | + | + | + | + | + | + | + | + | Leu |
| Gln-143 | + | + | + | + | + | + | + | + | + |
| Leu-142 | + | Phe | Phe | Phe | + | + | + | + | + |
| Ile-141 | + | + | + | + | + | + | + | + | + |
| Leu-140 | + | + | + | + | Met | + | + | + | + |
| Asp-139 | + | Asn | Asn | Asn | Asn | Asn | Asn | Asn | Asn |

*Table 3.2* continuation

| Con A | DL | PsL | LcL | VfL | SBA | PHA-L | PHA-E | PHA-P | SL |
|---|---|---|---|---|---|---|---|---|---|

β-pleated sheets (residues of Con A back sheet):

| Con A | DL | PsL | LcL | VfL | SBA | PHA-L | PHA-E | PHA-P | SL |
|---|---|---|---|---|---|---|---|---|---|
| Thr-147 | Phe | Tyr | Tyr | Tyr | Ile | Ser | + | Ser | Val |
| Thr-148 | + | + | + | + | Val | Val | Val | Val | + |
| - | - | - | - | - | - | Ser | Ser | Ser | Asp |
| Gly-149 | Asp | - | - | - | Thr | - | - | - | Asp |
| Leu-156 | + | + | + | + | + | + | + | + | + |
| Glu-155 | + | Thr | Thr | Thr | Gln | Arg | Arg | Arg | Val |
| Leu-154 | + | + | + | + | + | + | + | + | + |
| Asn-153 | + | Lys | Gly | Lys | Lys | Gln | Gln | Gln | Cys |
| Gly-152 | + | Glu | Glu | Glu | + | + | + | + | Arg |
| Phe-130 | + | Ile | Ile | Ile | Trp | Arg | Arg | Arg | + |
| Met-129 | Ser | Leu | Ser | Ser | Ser | Gln | Gln | Asp | Asp |
| Phe-128 | + | + | + | + | + | + | + | + | + |
| His-127 | + | - | - | - | - | - | - | - | - |
| Leu-126 | + | Ser | Ser | Ser | Ser | Asn | Ser | Ser | Ser |
| Ala-125 | Ser | Thr | Thr | Thr | Val | Phe | Phe | Phe | Val |
| Asp-124 | Asn | - | - | - | - | Tyr | Ser | Phe | - |
| Thr-123 | Ala/Glu | + | + | Ile | + | Ile | + | + | + |
| Thr-103 | + | Ala | Ala | + | Gly | Val | Val | Val | Gln |
| Asn-104 | + | His | Gln | His | Glu | Glu | Glu | Glu | His |
| Thr-105 | + | Glu | Glu | Glu | Ser | + | + | + | Arg |
| Ile-106 | + | Val | Val | Val | His | Asn | Asn | Asn | Leu |
| Ser-108 | + | + | + | + | + | + | + | + | + |
| Trp-109 | + | + | + | + | + | + | + | + | + |
| Ser-110 | + | + | + | Thr | + | + | + | + | + |
| Phe-111 | + | + | + | + | + | + | + | + | + |
| Thr-112 | + | His | Asn | Leu | Ala | Ala | Ala | Ala | Lys |
| Ser-113 | + | + | + | + | + | + | + | + | + |
| Lys-114 | + | Glu | Gln | Gln | Asn | + | + | + | Val |
| Leu-115 | + | + | + | + | + | + | + | + | + |
| Lys-116 | + | Ser | Gly | Thr | Pro | Ser | Ser | Ser | Pro |
| Lys-200 | + | Asn | Arg | Asp | Tyr | Gln | Asp | Leu | Tyr |
| Ile-199 | + | + | Phe | + | Phe | + | + | + | + |
| Leu-198 | + | Val | Val | Val | Thr | Asn | Asn | Asn | Phe |
| Phe-197 | + | + | Gln | + | + | + | + | + | + |
| Ala-196 | Ile/Tyr | Thr | Ser | Ile | Asn | Thr | Thr | Thr | Thr |
| Phe-195 | + | + | Gly | + | + | + | + | + | + |
| Thr-194 | + | Ser | Asn | + | Ser | Ser | Ser | Asn | Ser |
| Ala-193 | + | Thr | Thr | Thr | + | Thr | Thr | Thr | Thr |
| Glu-192 | Asp | Val | Val | Thr | Ala | Ala | Pro | Asp | + |
| Phe-191 | + | + | + | + | + | + | Ser | + | + |
| Ala-190 | Ser | Asn | Asn | Asp | Ser | Ser | + | Ser | Ser |
| Ser-189 | Ala | Ala | Ala | Ala | Ala | Ala | Ala | Ala | Ala |
| Val-188 | + | + | + | + | + | + | + | + | Glu |
| Gly- 48 | + | Ala | - | Ala | Ala | Ala | Ala | Ala | - |
| Thr- 49 | + | Asn | - | His | Lys | Glu | Glu | Glu | + |
| Ala- 50 | Val | Val | - | Val | Val | Val | Val | Val | Val |
| His- 51 | + | Val | - | Ala | Leu | Leu | Leu | + | Thr |
| Ile- 52 | + | + | - | + | + | + | + | + | + |
| Ile- 53 | Ser | Ala | - | Ser | Thr | Thr | Thr | Thr | Thr |
| Tyr- 54 | + | Phe | - | Phe | + | + | + | + | + |

*Table 3.2* continuation

| Con A | DL | PsL | LcL | VfL | SBA | PHA–L | PHA–E | PHA–P | SL |
|-------|----|-----|-----|-----|-----|-------|-------|-------|----|
| Asn– 55 | + | + | – | + | Asp | Asp | Asp | Glu | Asp |
| Ser– 56 | + | Ala | – | Ala | Ala | + | + | + | Ala |
| Ser– 66 | + | Thr | – | Leu | Pro | Pro | Pro | Pro | Asn |
| Val– 65 | + | Leu | – | Leu | Tyr | Tyr | Tyr | Tyr | Arg |
| Val– 64 | + | Ser | – | Thr | + | + | + | + | Tyr |
| Ala– 63 | + | Val | – | Val | Leu | Leu | Leu | Leu | Phe |
| Ser– 62 | + | Thr | – | + | + | + | + | + | + |
| Leu– 61 | + | + | – | + | Ala | Ala | Ala | Ala | Ser |
| Arg– 60 | + | Val | – | Val | Val | Val | Val | Val | Val |
| Lys– 59 | + | Asn | – | Asn | Leu | Leu | Leu | Leu | Ser |
| Thr– 73 | + | Tyr | Tyr | Tyr | Asn | Phe | Phe | Phe | Phe |
| Ser– 74 | Thr | Thr | Thr | Thr | Ile | Ile | Ile | Thr | Thr |
| Val– 75 | + | Leu | Leu | Leu | Leu | + | + | + | + |
| Ser– 76 | + | + | Asn | + | + | + | + | + | Lys |
| Tyr– 77 | + | Asp | Glu | Glu | Asp | Asp | Asp | Asp | Ala |
| Asp– 78 | + | Val | Val | Val | Val | Thr | Thr | Thr | Ser |
| Val– 79 | + | + | + | + | + | + | + | + | + |

**Residues of Con A α-helix-like structure:**

| Con A | DL | PsL | LcL | VfL | SBA | PHA–L | PHA–E | PHA–P | SL |
|-------|----|-----|-----|-----|-----|-------|-------|-------|----|
| Asp– 80 | + | Ser | Pro | Pro | + | + | + | + | His |
| Leu– 81 | + | + | + | + | + | + | + | + | + |
| Asn– 82 | + | Lys | Lys | Lys | Lys | Lys | Lys | Lys | Arg |
| Asp– 83 | Asn | + | + | + | Thr | Ser | Ser | Ser | + |
| Val– 84 | + | + | + | + | Ser | + | + | + | Ala |
| Leu– 85 | + | Val | Val | Val | + | + | + | + | + |

## 3.2.3 In Vivo Biosynthesis and Transport of Legume Lectins

Besides Con A, lectins have been purified and characterized mainly from the seeds of many Leguminosae (Table 3.3, see pp. 94–98). On the basis of the molecular weight of their subunits, legume lectins are generally classified into two groups, i.e., the one-chain and the two-chain lectins. The former group of lectins are either dimers or tetramers of one type of subunit, while the latter are built up of two different subunits which are generally denoted as α- or light chain, and β- or heavy chain. Both chains are held together by noncovalent forces.

For several two-chain lectins it has been well established that both chains are not the product of separate genes, but rather result from the proteolytic cleavage of a precursor chain. This notion is supported by several lines of evidence. In their study on the biosynthesis of favin, Hemperly et al. (1982) demonstrated that, upon in vitro translation and immunoprecipitation of the translated products with antiserum directed against the lectin, one single polypeptide of molecular weight 29,000 had been synthesized. This polypeptide was shown by sequence analysis to consist of a stretch of 29 hydrophobic amino acids, resembling the signal peptides encountered in several animal and prokaryotic precursor proteins (Blobel and Dobberstein 1975; Walter and Blobel 1981), followed by a sequence which is identical to that of the β-chain of mature favin. Hemperly et al. (1982) could also prove that the α-chain sequence is included in the 29,000 MW

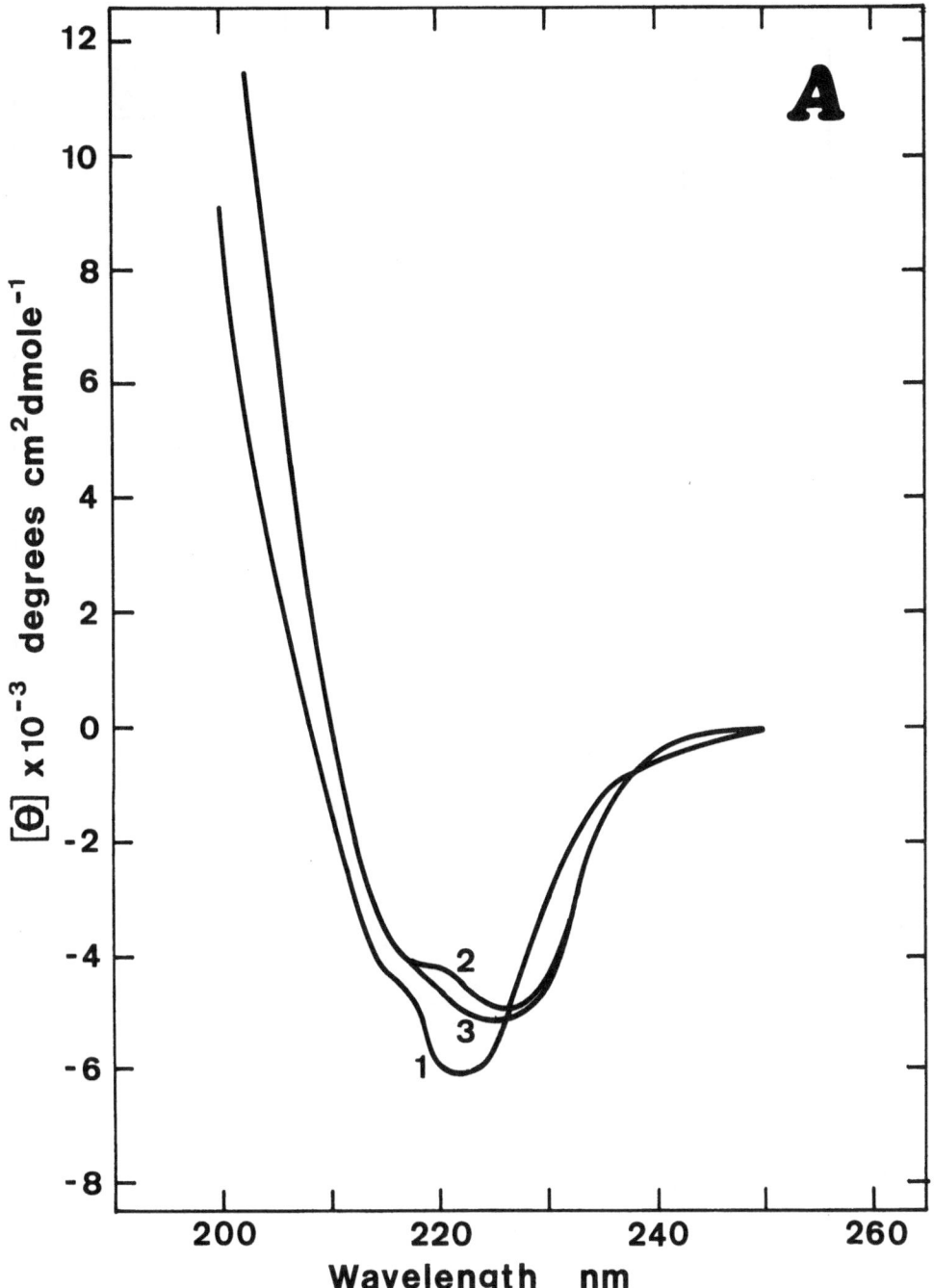

*Fig. 3.4* Circular dichroism spectra

A, B) in the far-UV region (195–250 nm) of concanavalin A (1), and the seed lectins of Pisum sativum (2), Lens culinaris (3), Vicia sativa (4) and Vicia faba (5)

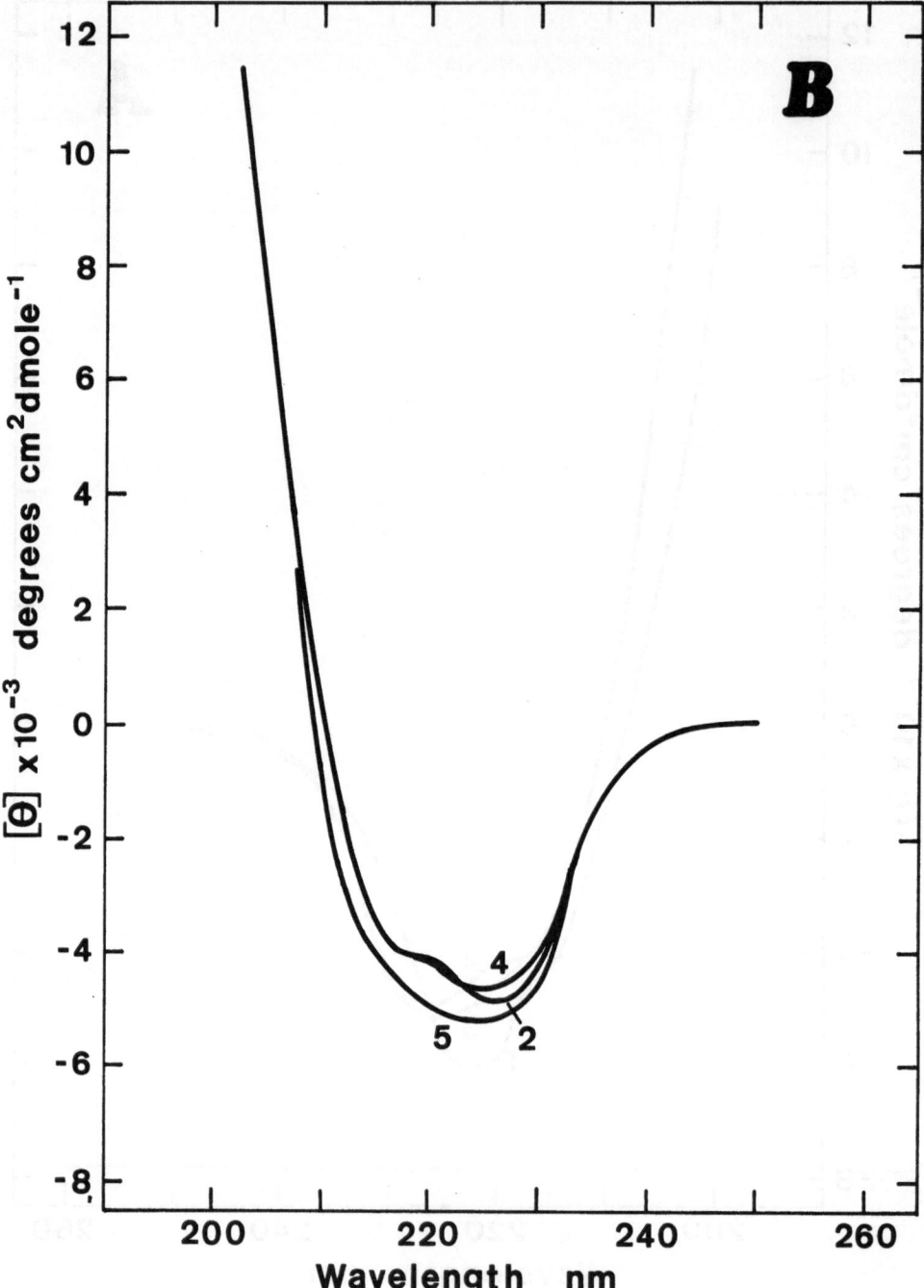

*Fig. 3.4* continuation

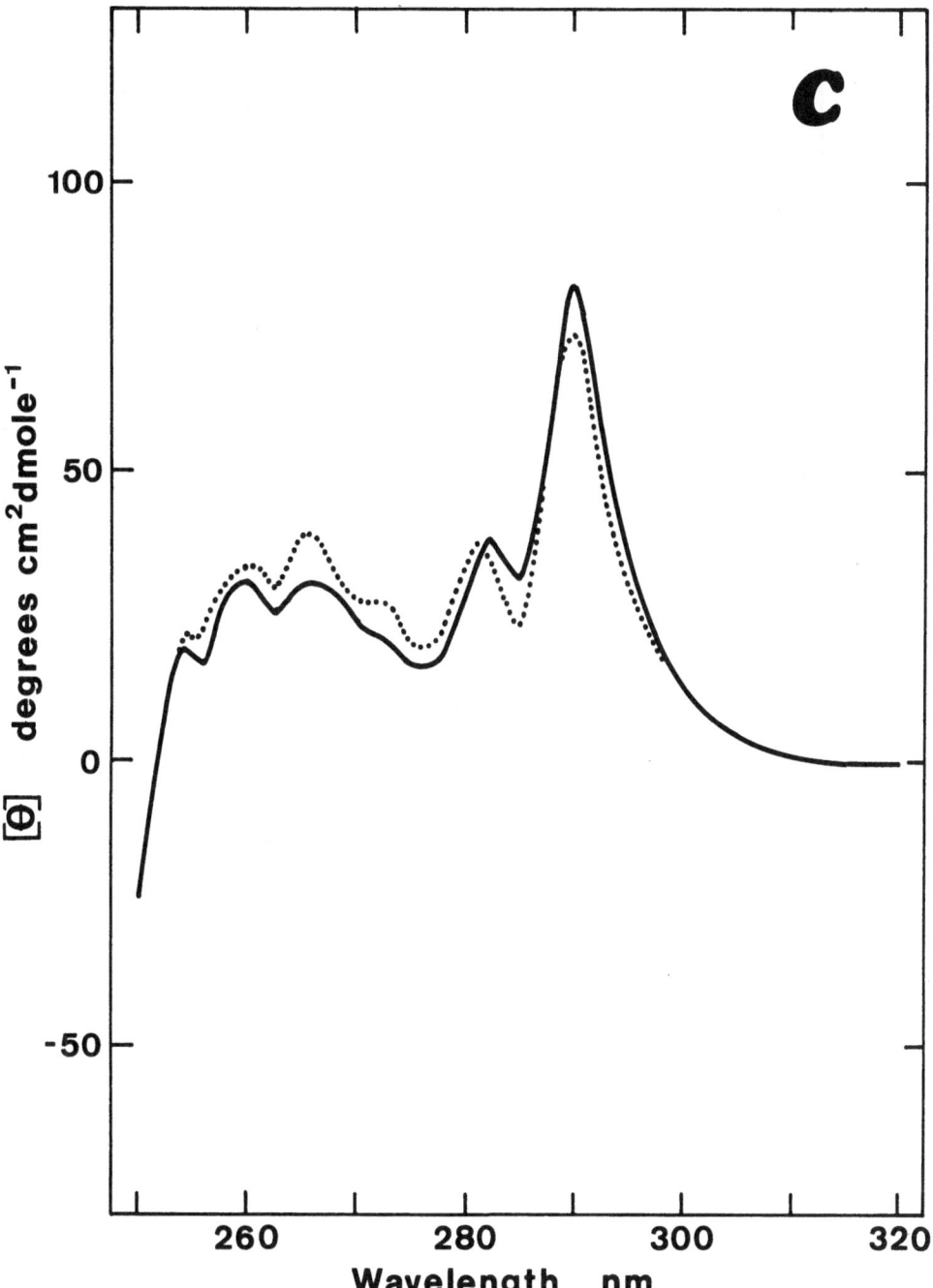

*Fig. 3.4* continuation

C–G) in the near-UV region (250–320 nm) of concanavalin A (C), and the seed lectins of Pisum sativum (D), Lens culinaris (E), Vicia sativa (F) and Vicia faba (G) (full lines)

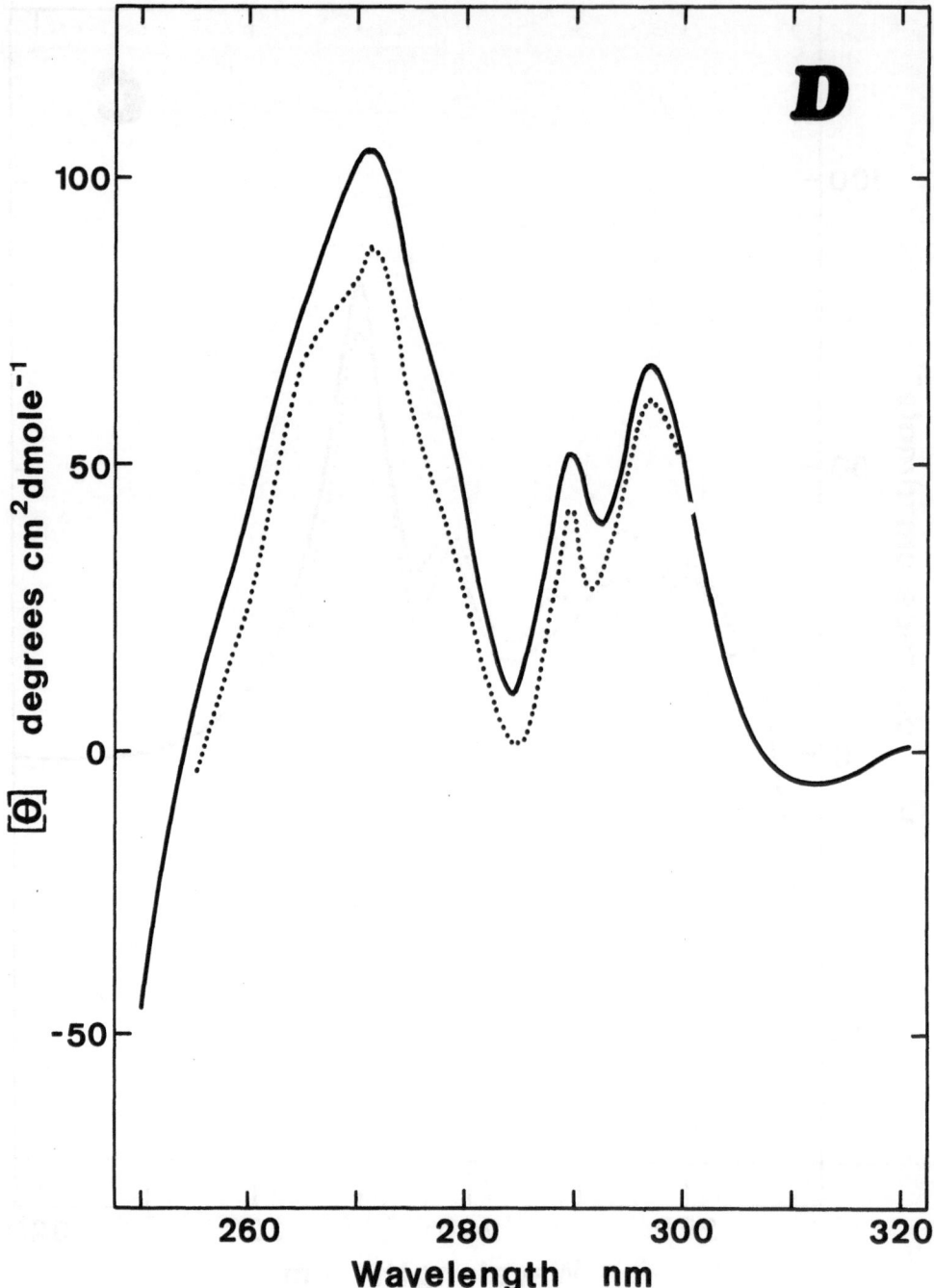

*Fig. 3.4*   continuation

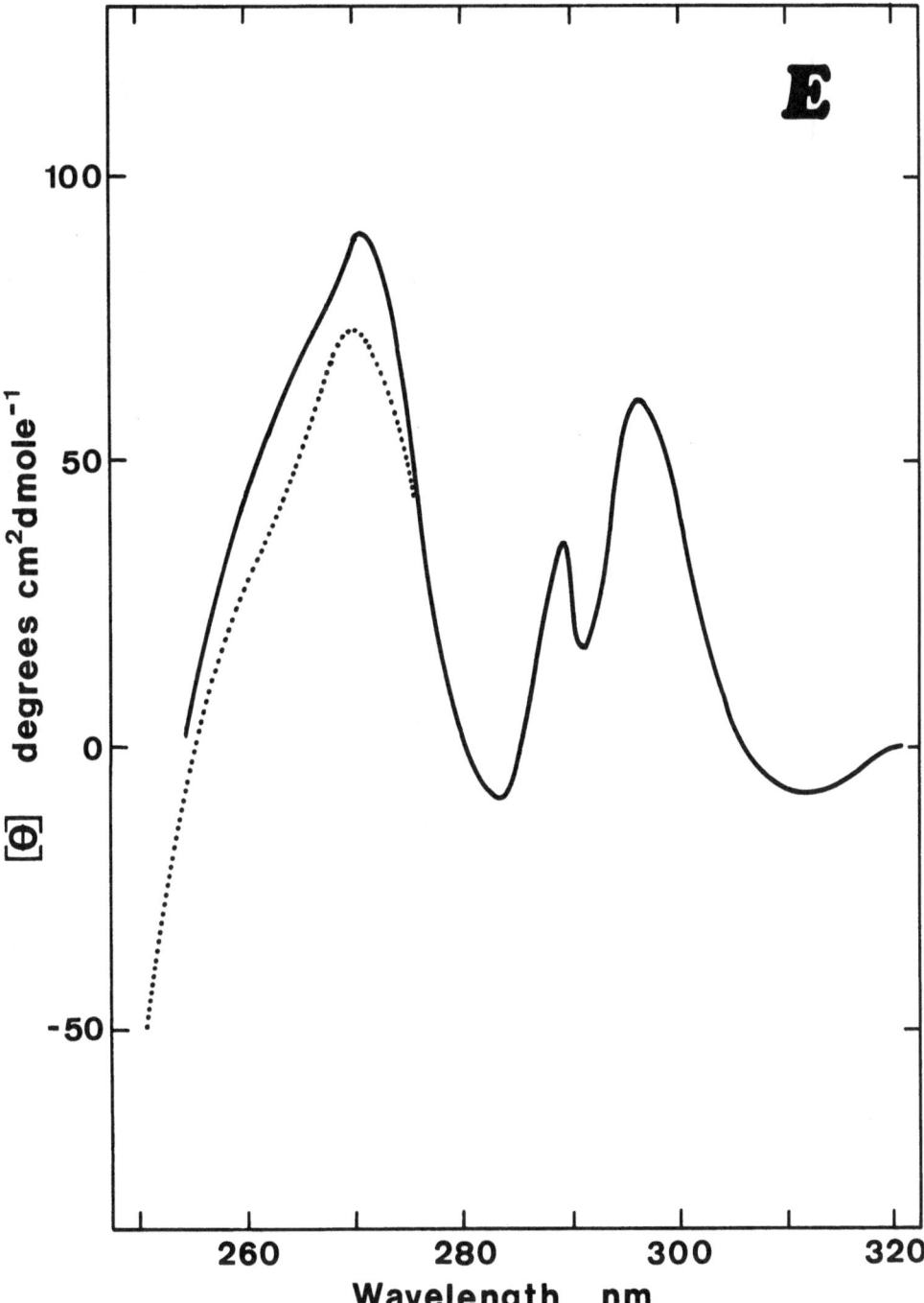

*Fig. 3.4* continuation

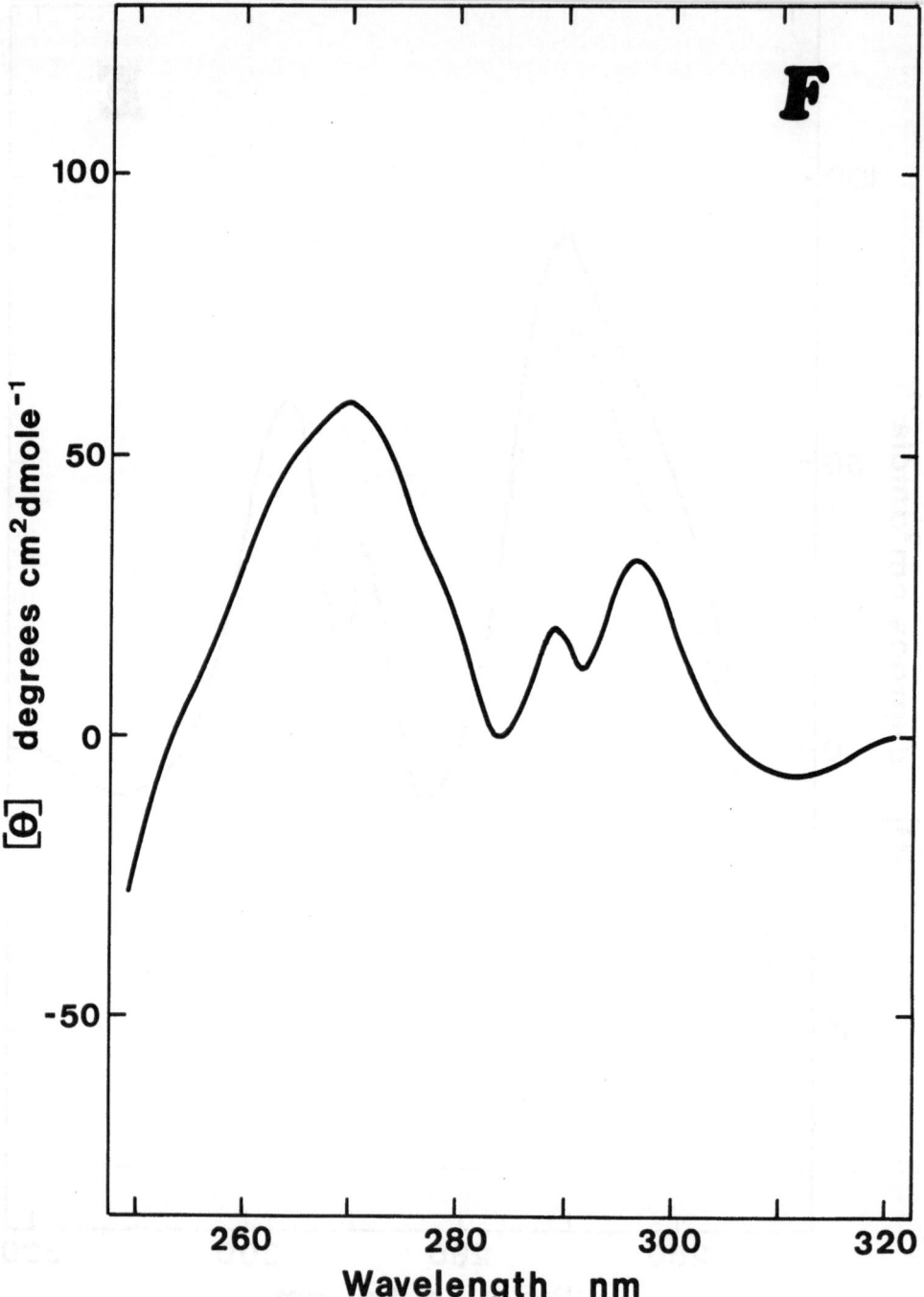

*Fig. 3.4* continuation

92

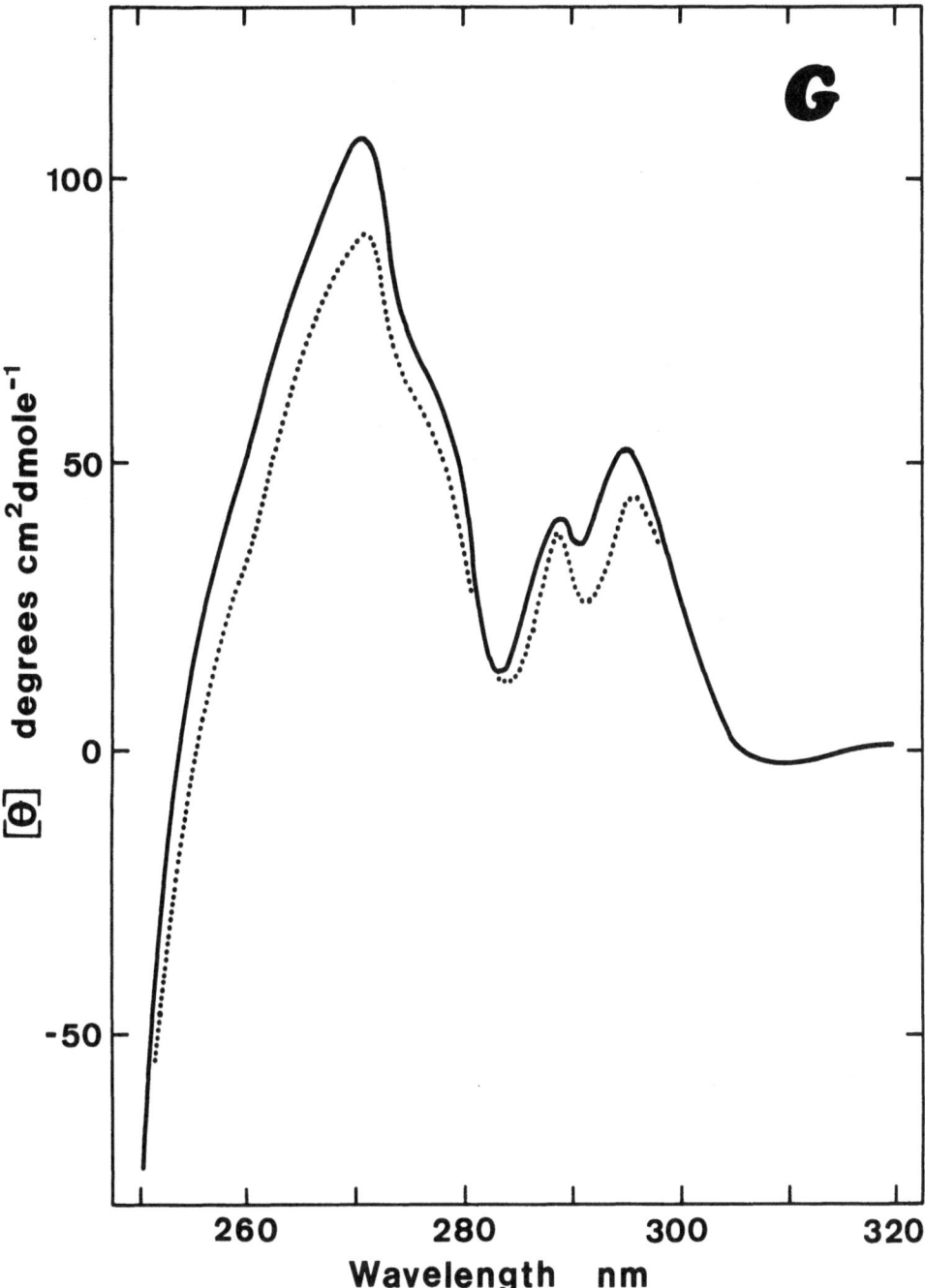

*Fig. 3.4* continuation

Lectins were prepared and stored as described by van Driessche et al. (1982). Circular dichroism spectra were recorded on a Cary 61 spectropolarimeter. The lectins were dissolved in 1 M NaCl, containing 10 mM potassium phosphate, pH 7.2. A structural relatedness of legume lectins is evident. From these figures, as well as from sequence data, it is obvious that the Viciae lectins are more related to each other than to concanavalin A. The addition of methyl-α-D-mannopyranoside (at a final concentration of 10 mM) induces a conformational change in all these lectins except in Vicia sativa (C–G, dotted lines).

Table 3.3 Physico-chemical properties of leguminosae lectins

| Lectin | MW (kd) | Subunits MW (kd) | Number | Carbohydrate content | Specificity Sugar | Human blood | Metal requirem. | Ref. |
|---|---|---|---|---|---|---|---|---|
| **I. SUBFAMILY: Papilionaceae** | | | | | | | | |
| **1. Tribe: Carageae** | | | | | | | | |
| 1.1. *Caragana arborescens* I | 120 | 29.4–30.4 | 4 | yes | GalNAc | none | nd. | 1 |
| II | 60 | 29–32 | 2 | yes | nd. | nd. | nd. | 1 |
| **2. Tribe: Cicereae** | | | | | | | | |
| 2.1. *Cicer arietinum* | 44 | 26 | 2 | nd. | no simple sugar fetuin, IgM | none | nd. | 2 |
| **3. Tribe: Diocleae** | | | | | | | | |
| 3.1. *Canavalia ensiformis* | 52 | 26 | 2 | none | mannose/glucose | none | Mn, Ca | 3–10 |
| | 104 | 26 | 4 | | | | | |
| 3.2. *Dioclea grandiflora* | 100 | 10–13 | (fragments) | none | mannose/glucose | nd. | Mn, Ca | 11, 12 |
| | | 26 ($\alpha$) | $\alpha_2\beta_2\gamma_2$ | | | | | |
| | | 13–14 ($\beta$) | | | | | | |
| | | 8–9 ($\gamma$) | | | | | | |
| | 50 | idem. | | | | | | |
| **4. Tribe: Galegeae** | | | | | | | | |
| 4.1. *Robinia pseudoacacia* | | | | | | | | |
| seeds | 100 | 30 | 4 | 17% | no simple sugar | nd. | nd. | 13 |
| seed RPA-I | 59 | 34 | 2 | 11.6% | no simple sugar | nd. | nd. | 14 |
| | 110 | 34 | 4 | 11.6% | no simple sugar | nd. | nd. | 15 |
| RPA-II | 105 | 27–30.5 | 4 | 4.3% | GalNAc | nd. | nd. | 14, 15 |
| | | 29 | | | | | | |
| bark | 110 | 29–31.5 | 4 | 7.5% | GalNAc | none | Mn, Zn | 16 |
| 4.2. *Wistaria floribunda* | 68–136 | 32–35 | 2–4 | 3.2% | GalNAc | none | none | 17–20 |

94

Table 3.3 continuation

| Lectin | MW (kd) | Subunits MW (kd) | Number | Carbo-hydrate content | Specificity Sugar | Human blood | Metal requirem. | Ref. |
|---|---|---|---|---|---|---|---|---|
| **5. Tribe: Genisteae** | | | | | | | | |
| 5.1. Crotolaria juncea | 120 | 31.4 | 4 | 9.8% | galactose | nd. | Mn, Ca | 21, 22 |
| 5.2. Cytisus scoparius CSIa | 70 | 30 | 2–4 | nd. | galactose/GalNAc | nd. | Mn, Ca, Mg | 23 |
| CSIb | 70 | 30 | 2–4 | nd. | GalNAc | nd. | nd. | 23 |
| CSII | 75 | 30 | 2–4 | nd. | GalNAc | nd. | Mn, Ca, Mg | 23 |
| 5.3. Cytisus sessilifolius | 110 | nd. | nd. | nd. | di-GlucNAc | nd. | nd. | 24, 25 |
| 5.4. Ulex europaeus I | 60–68 | 32 } 29 | 2 | 5.2–7.2% | L-fucose | O | Mn, Ca, Zn | 26–29 |
| II | 105 | 23–25 | 4 | 10.3% | di-GlucNAc | O | Mn, Ca | 29–32 |
| **6. Tribe: Hedysareae** | | | | | | | | |
| 6.1. Arachis hypogaea | 98–110 | 27–28 | 4 | none | galactose | none | Mn, Ca | 33–40 |
| 6.2. Onobrychis viciifolia seeds, roots | 52 | 26 | 2 | 4.2% | mannose/glucose | none | Mn, Ca | 41–45 |
| **7. Tribe: Loteae** | | | | | | | | |
| 7.1. Lotus tetragonolobus A | 100 | 27.8 | 4 | 8% | L-fucose | O | nd. | 46–48 |
| B | 58 | 27.0 | 2 | 4% | L-fucose | O | nd. | 46–48 |
| C | 117 | 27.8 | 4 | 8% | L-fucose | O | nd. | 46–48 |
| **8. Tribe: Phaseoleae** | | | | | | | | |
| 8.1. Amphicarpaea bracteata | 135 | 32, 30, 28, 27.5 } | 4 | 2% | GalNAc | $A_1$ | nd. | 49–50 |
| 8.2. Dolichos biflorus A | 113 | 26.5 | 4 | 2% | GalNAc | A | Ca | 51–56 |
| B | 109 | 26 | 4 | yes | GalNAc | A | Ca | 51–56 |
| CRM | 68–70 | 34–35 | 2 | yes | GalNAc/GlucNAc | none | nd. | 57, 58 |
| 8.3. Erythrina cristagalli | 56.8 | 26–28 | 2 | 4.5% | galactose | none | Mn, Ca | 59 |

*Table 3.3* continuation

| Lectin | MW (kd) | Subunits MW (kd) | Number | Carbohydrate content | Specificity Sugar | Human blood | Metal requirem. | Ref. |
|---|---|---|---|---|---|---|---|---|
| 8.4. *Glycine max* | 120 | 30 | 4 | 5.5% | GalNAc | none | Mn, Ca | 60–65 |
| | 175 | 45 | 4 | 3–5% | 4-o-methyl-D-glucuronic acid | nd. | nd. | 66, 67 |
| 8.5. *Phaseolus lunatus* II | 247.1 | 31 | $(\alpha_2)_4$ | 3–4% | GalNAc | A | Mn, Ca | 68–72 |
| III | 124.4 | 31 | $(\alpha_2)_2$ | 3–4% | GalNAc | A | Mn, Ca | 68–72 |
| 8.6. *Phaseolus vulgaris* | 136 | L: 34 E: 34 | $L_4$ $L_3E_1$ $L_2E_2$ $L_1E_3$ $E_4$ | 6.2% | L: no simple sugar E: no simple sugar | pref. A | Mn, Ca | 73–80 |
| cultivar Pinto III | 52.2–55 | 28–29 | 2 | 12–19% | no simple sugar | nd. | nd. | 81, 82 |
| 8.7. *Vigna radiata* | 160 | 45.8 | 4 | nd. | galactose | none | nd. | 83, 84 |
| **9. Tribe: Sophoreae** | | | | | | | | |
| 9.1. *Sophora japonica* | 132.8 | 32.5 | 4 | 8% | GalNAc | pref. B | Mn, Ca | 85–87 |
| **10. Tribe: Trifolieae** | | | | | | | | |
| 10.1. *Ononis hircina* | 110 | 31 | 4 | 7.2% | GalNAc | pref. O | nd. | 88 |
| 10.2. *Trifolium repens* | nd. | 50 | nd. | yes | 2-deoxy-glucose | none | Mn, Ca, Mg | 89 |
| **11. Tribe: Viciae** | | | | | | | | |
| 11.1. *Abrus precatorius* | 126–135 | 33 ($\alpha$) 36 ($\beta$) 37.5 ($\beta'$) | 2 1 1 | 5% | galactose/lactose | none | none | 90–92 |
| 11.2. *Lathyrus ochrus* | 46 | 17 ($\beta$) 6 ($\alpha$) | 2 2 | 0.25% | mannose/glucose | none | Mn, Ca | 93 |
| 11.3. *Lathyrus odoratus* | 50 | 19 ($\beta$) 6 ($\alpha$) | 2 2 | 0.5% | mannose/glucose | none | Mn, Ca | 94–97 |

Table 3.3 continuation

| Lectin | MW (kd) | Subunits | | Carbohydrate content | Specificity | | Metal requirem. | Ref. |
|---|---|---|---|---|---|---|---|---|
| | | MW (kd) | Number | | Sugar | Human blood | | |
| 11.4. *Lathyrus sativus* | 43 | 21.5 (β)<br>10 (α)<br>or: $\beta\alpha_2$ | 2<br>2 | 2% | mannose/glucose | none | nd. | 98 |
| 11.5. *Lathyrus tingitanus* | 49 | 19 (β)<br>4.4 (α) | 2<br>2 | 0.5% | mannose/glucose | none | nd. | 99 |
| 11.6. *Lens culinaris* | 50 | 20 (β)<br>5 (α) | 2<br>2 | 0.25% | mannose/glucose | none | Mn, Ca | 100 |
| | 46.4 | 17.6 (β)<br>5.6 (α) | 2<br>2 | none | mannose/glucose | none | Mn, Ca | 101–109 |
| 11.7. *Pisum sativum* | 50 | 18 (β)<br>5.8 (α)<br>6.5 fragments<br>8.5.<br>26–28 precursor | 2<br>2 | none | mannose/glucose | none | Mn, Ca | 110–124 |
| 11.8. *Vicia cracca* | 125 | 33 | 4 | nd. | GalNAc | $A_1$ | nd. | 125 |
| | 44 | 17.5 (β)<br>5.7 (α) | 2<br>2 | nd. | mannose/glucose | none | nd. | 126 |
| 11.9. *Vicia ervilia* | 60 | 21 (β)<br>4.7 (α) | 2<br>2 | none | mannose/glucose | none | nd. | 127 |
| 11.10. *Vacia faba* | 51 | 20 (β)<br>2.6 (α)<br>9.5 fragments<br>11.6<br>29 precursor | 2<br>2 | 3% | mannose/glucose | none | nd. | 128–133 |
| 11.11. *Vicia graminea* | 100 | 25 | 4 | 7.3% | no simple sugars tryptic human erythrocyte pept. | N | nd. | 134, 135 |
| | 125 | 31 | 4 | nd. | no simple sugars | N | nd. | 136 |

Table 3.3 continuation

| Lectin | MW (kd) | Subunits MW (kd) | Subunits Number | Carbohydrate content | Specificity Sugar | Specificity Human blood | Metal requirem. | Ref. |
|---|---|---|---|---|---|---|---|---|
| 11.12. *Vicia hirsuta* | nd. | 19.2 (β) 12.8 (α) | nd. | none | mannose/glucose | none | nd. | 137 |
| 11.13. *Vicia sativa* | 70 | 27.5 20 | nd. | 9.4% | mannose/glucose | none | nd. | 138 |
|  | 40 | 14 (β) 5–6 (α) | 2 2 | nd. | mannose/glucose | pref. B | nd. | 139, 140 |
|  | nd. | 20.5 (β) 23 (β') 5.8 (α) | nd. | nd. | mannose/glucose | nd. | nd. | 141 |
| 11.14. *Vicia villosa* | 134.4–143.6 | 35.9 (β) 33.6 (α) | 2 2 or: $\alpha_4, \beta_4$ | 7–10% | GalNAc | $A_1, A_4$ | nd. | 142 |
|  | 120 | 30 | 4 | 4.3% | GalNAc | $A_1$ | Mn, Zn | 143–144 |
| **II  SUBFAMILY: Caesalpinoideae** |  |  |  |  |  |  |  |  |
| 1. *Bauhinia purpurea* | 120 | 32 | 4 | 11–18% | GalNAc | none | nd. | 145, 146 |
| 2. *Bandeiraea simplicifolia* |  |  |  |  |  |  |  |  |
| I | 114 | A: 32 B: 30 | 4 4 | 6% 6% | GalNAc/galactose Gal | A B | Ca Ca | 147–152 147–152 |
| II | 120 | 30 | 4 | 4% | GlucNAc | none | nd. | 153 |

For more details on the physico-chemical characteristics of the lectins mentioned, the reader should consult the review written by Goldstein and Poretz (1986).

Abbreviations used:
MW: molecular weight    CRM: cross reacting material    GalNAc: N-acetyl-galactosamine; GlucNAc: N-acetyl-glucosamine
nd.: not determined    pref.: preferentially

*Table 3.3* continuation

Ref. – references:

**1.** Bloch et al. 1976; **2.** Kolberg et al. 1983; **3.** Wang et al. 1971; **4.** Cunningham et al. 1972; **5.** Shoham et al. 1973; **6.** Wang et al. 1975; **7.** Cunningham et al. 1975; **8.** Blumberg and Tal 1976; **9.** Shoham et al. 1978; **10.** Reeke et al. 1978; **11.** Moreira et al. 1983; **12.** Richardson et al. 1985a; **13.** Bourillon and Font 1968; **14.** Wantyghem et al. 1986; **15.** Fleischmann and Rüdiger 1986; **16.** Horejsi et al. 1978b; **17.** Toyoshima and Osawa 1975; **18.** Kurokawa et al. 1976; **19.** Cheung et al. 1979; **20.** Sugii and Kabat 1980; **21.** Ersson et al. 1973; **22.** Ersson 1976; **23.** Young et al 1984; **24.** Matsumoto and Osawa 1972; **25.** Matsumoto and Osawa 1974; **26.** Matsumoto and Osawa 1969; **27.** Horejsi and Kocourek 1974; **28.** Frost et al. 1975; **29.** Osawa and Matsumoto 1972; **30.** Matsumoto and Osawa 1970; **31.** Pereira et al. 1979; **32.** Konami et al. 1981; **33.** Lotan et al. 1975b; **34.** Terao et al. 1975; **35.** Baues and Gray 1977; **36.** Sutoh et al. 1977; **37.** Fish et al. 1978; **38.** Neurohr et al. 1980; **39.** Miller 1983; **40.** Uy and Wold 1977; **41.** Hapner and Robbins 1979; **42.** Namen and Hapner 1979; **43.** Young et al. 1982; **44.** Kouchalakos and Hapner 1984; **45.** Kouchalakos et al. 1984; **46.** Yariv et al. 1967; **47.** Kalb 1968; **48.** Pereira and Kabat 1974; **49.** Blacik et al. 1978; **50.** Goldstein and Poretz 1986; **51.** Etzler and Kabat 1970; **52.** Carter and Etzler 1975; **53.** Hammarström et al. 1977; **54.** Etzler et. 1977; **55.** Borrebaeck et al. 1981; **56.** Etzler et al. 1981; **57.** Talbot and Etzler 1978; **58.** Etzler and Borrebaeck 1980; **59.** Iglesias et al. 1982; **60.** Lis et al. 1966a; **61.** Lis et al. 1966b; **62.** Catsimpoolas and Meyer 1969; **63.** Jeffe et al. 1977; **64.** Lis and Sharon 1978; **65.** Vodkin et al. 1983; **66.** Dombrink-Kurtzman et al. 1983; **67.** Rutherford et al. 1986; **68.** Galbraith and Goldstein 1970; **69.** Galbraith and Goldstein 1972; **70.** Pandolfino and Magnuson 1980; **71.** Roberts and Goldstein 1984; **72.** Nissen and Magnuson 1986; **73.** Miller et al. 1973; **74.** Pusztai and Watt 1974; **75.** Miller et al. 1975; **76.** Felsted et al. 1977; **77.** Dupuis and Leclair 1982; **78.** Hammerström et al. 1982; **79.** Hoffman and Donaldson 1985; **80.** Yamashita et al. 1983; **81.** Pusztai et al. 1981; **82.** Pusztai et al. 1982; **83.** Hankins and Shannon 1978; **84.** Del Campillo et al. 1981; **85.** Poretz et al. 1974; **86.** Timberlake et al. 1980; **87.** Wu et al. 1981; **88.** Horejsi et al. 1978a; **89.** Dazzo et al. 1978; **90.** Olsnes et al. 1974; **91.** Wei et al. 1975; **92.** Roy et al. 1976; **93.** Richardson et al.1984b; **94.** Kolberg and Michaelsen 1979; **95.** Kolberg et al. 1980; **96.** Etzler et al. 1981; **97.** Sletten et al. 1983; **98.** Gupta et al. 1980; **99.** Kolberg and Sletten 1982; **100.** Rougé and Chabert 1983; **101.** Howard et al. 1971; **102.** Paulova et al. 1971b; **103.** Fliegerova et al. 1974; **104.** Foriers et al. 1978; **105.** Foriers et al. 1977; **106.** Foriers et al. 1981; **107.** Kornfeld et al. 1981; **108.** Bhattacharyya and Brewer 1984; **109.** Bhattacharyya and Brewer 1985; **110.** Entlicher and Kocourek 1970; **111.** Paulova et al. 1971a; **112.** Marik et al. 1974; **113.** Trowbridge 1974; **114.** Van Wauwe et al. 1975; **115.** Van Driessche et al. 1976a; **116.** Van Driessche et. al. 1976b; **117.** Van Driessche et al. 1978; **118.** Richardson et al. 1978; **119.** Kornfeld et al. 1981; **120.** Van Driessche et al. 1982; **121.** Van Driessche et al. 1983; **122.** Higgins et al. 1983a; **123.** Higgins et al. 1983b; **124.** Bhattacharyya and Brewer 1985; **125.** Rüdiger 1977; **126.** Bauman et al. 1982; **127.** Fornstedt and Porath 1975; **128.** Allen et al. 1976; **129.** Horstmann et al. 1978; **130.** Hemperly et al. 1979; **131.** Cunningham et al. 1979; **132.** Hopp et al. 1982; **133.** Hemperly et al. 1982; **134.** Prigent and Bourillon 1976; **135.** Prigent and Bourillon 1980; **136.** Duk and Lisowska 1981; **137.** Solheim 1983; **138.** Falasca et al. 1979; **139.** Gebauer et al. 1979; **140.** Gebauer et al. 1981; **141.** Van Driessche et al. 1980; **142.** Tollefsen and Kornfeld 1983; **143.** Kaladas et al. 1981; **144.** Grubhoffer et al. 1981; **145.** Irimura and Osawa 1972; **146.** Young et al. 1985; **147.** Hayes and Goldstein 1974; **148.** Murphy and Goldstein, 1977; **149.** Wood et al. 1979; **150.** Murphy and Goldstein 1979; **151.** Delmotte and Goldstein 1980; **152.** Lamb et al. 1981; **153.** Iyer et al. 1976

99

polypeptide by isolating a peptide from the translation product which was shown by sequence determination to consist of residues 14−21 of the favin α-chain. Translation of favin mRNA in the presence of dog pancreas microsomal membranes revealed that the synthesized polypeptides are translocated into the lumen and glycosylated (Hemperly et al. 1982).

Independently, van Driessche et al. (1983) and Lauwereys et al. (1983) on the one hand, and Higgins et al. (1983a, 1983b) on the other hand could show that the two-chain lectin from pea seeds too is initially synthesized as a precursor chain. The idea that the subunits of pea lectin might be derived from a precursor chain had already been mentioned earlier (van Driessche et al. 1976b, 1978, 1982). These authors had noticed that upon SDS-electrophoresis of affinity purified pea lectin isolated from mature seeds, a minor component with a molecular weight of 26−28,000 was consistently present. This high molecular weight polypeptide was supposed to be the precursor chain. Unfortunately, attempts to purify this presumed precursor by gel filtration on Biogel P100, either in 6 M guanidine-HCl or in 6 M urea pH 3, was not quite satisfying due to the aggregation of the subunits in these conditions (van Driessche et al. 1976a). Protein fractions recovered from this gel filtration column were shown by SDS-electrophoresis to be highly enriched in precursor chains, but nevertheless always contaminated with β-chains (van Driessche et al. 1982). Automated sequence analysis of these fractions unequivocally revealed one unique sequence which was identical with that of highly purified β-chains (van Driessche et al. 1976a), pointing to a β-α-sequence in the precursor. Due to the β-chain contamination, at that time, the possibility could however not be excluded that the $NH_2$-terminal amino acid of the precursor was blocked. Attempts to purify the precursor chain from mature seeds by isoelectric focusing in granulated gels were more successful from the point of view of purity, but the quantities recovered were to small to allow further physico-chemical characterization.

Later it was noticed (van Driessche et al. 1983; Lauwereys et al. 1983; Higgins et al. 1983a) that appreciable amounts of precursor chains are present in immature seeds. From affinity purified pea lectin isolated from immature seeds, Lauwereys et al. (1983) succeeded in preparing a mixture of precursor- and β-chains which was completely devoid of α-chains. Digestion of this mixture with trypsin, and separation and sequencing of the purified peptides, revealed the presence of β-chain peptides as well as at least one internal peptide from the α-chain, indicating that the precursor chain contains both the β- and the α-sequence. More convincing evidence for this conclusion was presented (van Driessche et al. 1983; van Driessche et al., in preparation) after the pea lectin precursor had been purified by preparative isoelectric focusing using a 3−10/5−8 gradient in a granulated gel matrix. The results on the structural analysis of the pea lectin precursor were confirmed and extended by sequence analysis of cDNA plasmids derived from the lectin mRNA (Higgins et al. 1983a). These authors showed that the cDNA plasmids encode for both the β- and the α-subunits in a β→α orientation, as well as for a leader sequence of 124 nucleotides. Since no stop codon was found between the coding sequences, these authors concluded that pea lectin is synthesized as a pre-prolectin.

Studies on the intracellular sites of synthesis and processing of pea lectin in developing pea cotyledons (Higgins et al. 1983b) revealed that the protein is synthesized in association with the endoplasmatic reticulum (ER). Within the ER, the only processing step to occur is the co-translational removal of the leader sequence. From the ER, the prolectin moves toward the protein bodies, where it is slowly processed to yield both the β- and the α-subunits. Unlike favin, pea lectin is not glycosylated due to the missing of the required Asn-X-Thr sequence (see Fig. 3.1).

Our current understanding of the biosynthesis and of the many and complex post-synthetic modifications which may occur before legume lectins are finally deposited in the protein bodies is mainly the result of the outstanding work on phytohemagglutinin (PHA) performed by Chrispeels and associates (Chrispeels 1983; Vitale and Chrispeels 1984; Vitale et al. 1984a, 1984b).

The polypeptides of PHA are synthesized by polysomes attached to the endoplasmatic reticulum. The synthesis is accompanied by a co-translational removal of the signal peptide, as well as by a co-translational glycosylation, yielding polypeptides which are substituted with two high-mannose type oligosaccharide chains (Vitale et al. 1984a). In the Golgi apparatus, some of the high-mannose side chains of PHA are modified by the removal of five to six mannosyl residues and the addition of fucosyl and terminal N-acetylglucosamine residues (Vitale and Chrispeels 1984; Vitale et al. 1984b). From the Golgi-complex the lectin is transported to the protein bodies in small electron-dense vesicles which are derived from the Golgi apparatus (Chrispeels 1983). Upon arrival in the protein bodies, the modified oligosaccharide chains of PHA are gradually reduced in size by the removal of the terminal N-acetylglucosaminyl residues (Vitale and Chrispeels 1984).

Although Con A does not contain covalently attached carbohydrate, the protein is synthesized as a glycoprotein precursor (Herman et al. 1985) substituted with a high-mannose type oligosaccharide. It is mostly probable that the glycosylation site in Con A is located in the peptide consisting of 15 amino acid residues (see Fig. 3.3) which is removed upon proteolytic cleavage and ligation of pro-Con A to yield the mature subunit (Carrington et al. 1985).

The several processing steps which finally result in the conversion of the lectin precursor chains to the mature subunits are summarized in Figures 3.5 and 3.6. Whether the precursor chains will be cleaved or not seems to be determined by conformational restraints in the lectin precursor molecules, rather than by the lack of the processing enzyme(s) itself. This is clearly shown by the fact that in some seeds such as those of *Vicia cracca*, two lectins are present which are encoded by different genes (Baumann et al. 1979), i. e., a mannose/glucose specific two-chain lectin and a N-acetyl-galactosamine specific one-chain lectin.

Besides the mature subunits, some lectin preparations have been shown to contain natural fragments. Both in pea lectin (van Driessche et al. 1982; van Driessche et al., in preparation) and in favin (Hopp et al. 1982), the fragments result from a cleavage of the peptide bond between residues 76 and 77 of the β-chain. Besides, for favin (Hemperly et al. 1979) it has been noticed that the molecular weight heterogeneity of the β-chain is the result of the proteolytic removal of a peptide at the COOH-terminus of the β-chain and the concomitant loss of the covalently attached carbohydrate. Similarly, fragmented subunits have been reported to exist in Con A (see above) and soybean lectin preparations (Lotan et al. 1975a).

## 3.2.4    Isolectins, Pseudo-isolectins and Subunit Heterogeneity

Under appropriate conditions, some legume lectins can be separated into different molecular forms or isolectins. Unfortunately, only in a rather limited number of cases this molecular heterogeneity has been thoroughly investigated and explained.

A classical example of legume isolectins is that of phytohemagglutinin (PHA). PHA has been shown to consist of five tetrameric isolectins, designated $L_4$, $L_3E$, $L_2E_2$, $LE_3$ and $E_4$ (Allen et al. 1969; Miller et al. 1973, 1975; Pusztai and Watt 1974; Felsted et al. 1977; Leavitt et al. 1977; Monsigny et al. 1978). These isolectins can be separated from each other by ion-exchange chromatography on SP-Sephadex (Leavitt et al. 1977). The L- and E-subunits are encoded by different genes which were shown to be closely linked on the chromosome (Hoffman and Donaldson 1985). Despite the intensive sequence homology, the L- and E-subunits have clearly distinct biological properties (Leavitt et al. 1977) and differ in carbohydrate specificity (Cummings and Kornfeld, 1982a; Hammarström et al. 1982; Yamashita et al. 1983). $L_4$ is a po-

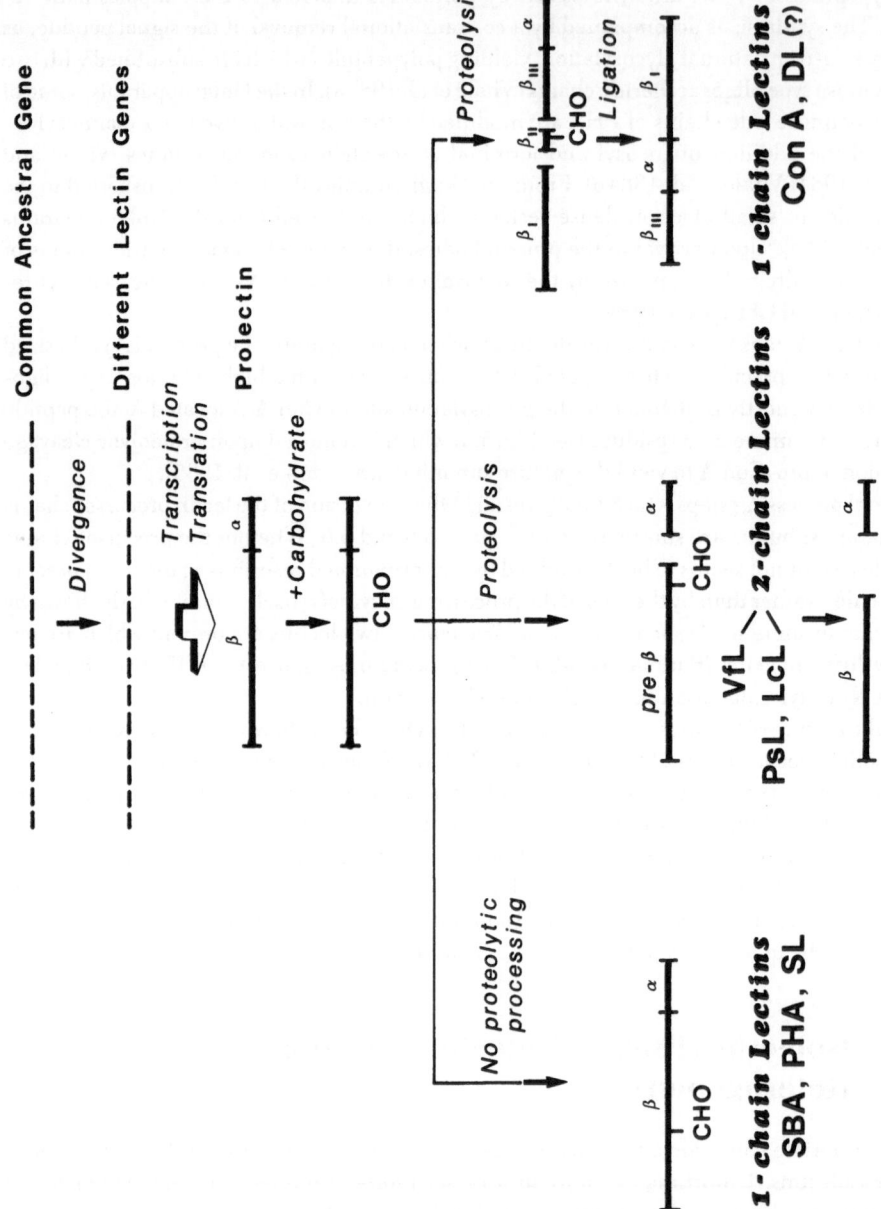

*Fig. 3.5* Ontogeny and processing modes of legume lectins
Abbreviations used are as mentioned in Figure 1.

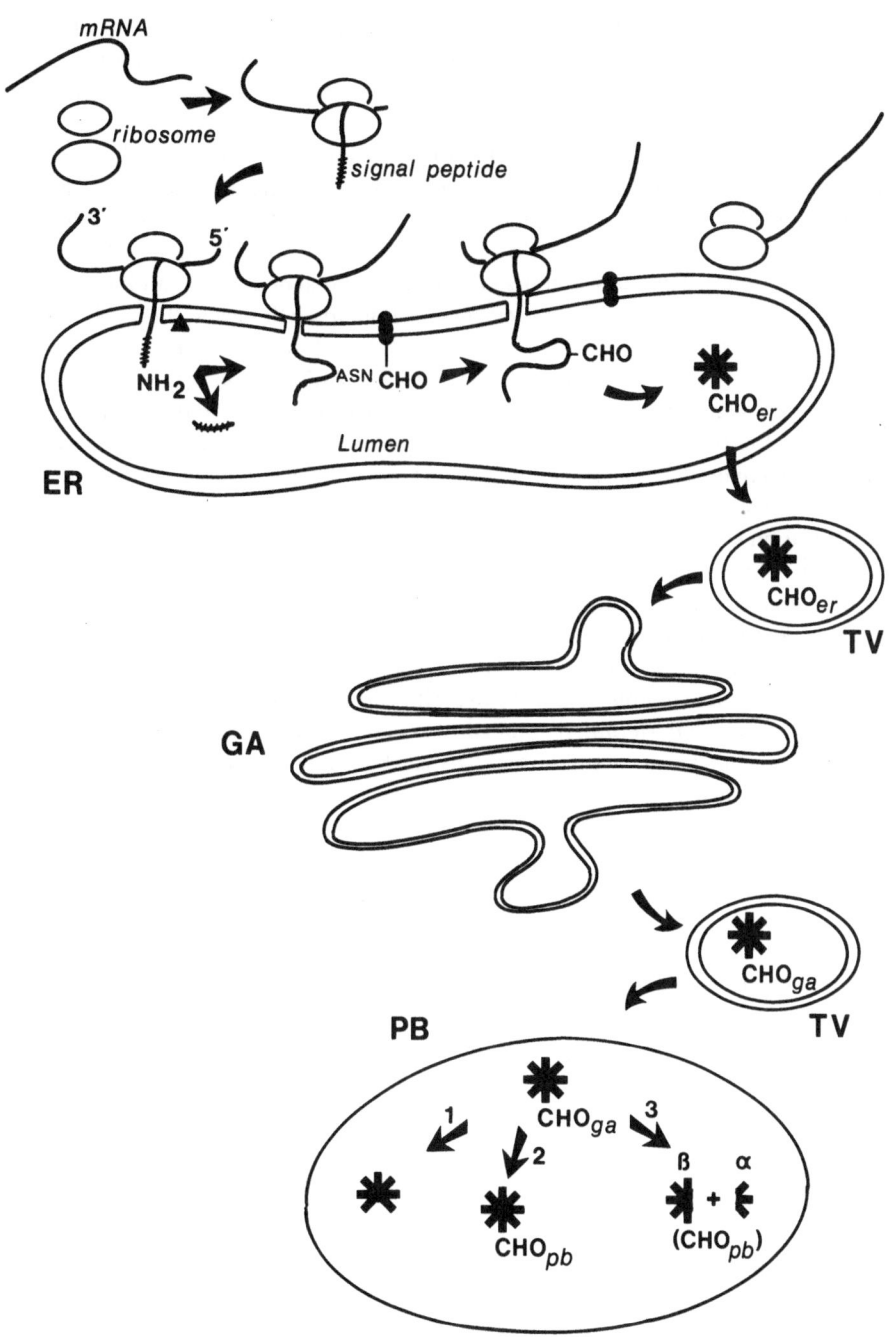

*Fig. 3.6* Intracellular sites of synthesis and processing of legume lectins

Abbreviations used:

| | |
|---|---|
| ER : endoplasmatic reticulum | CHO: oligosaccharide chains |
| TV : transport vesicle | ▲ : signal peptidase |
| GA: Golgi apparatus | ❚ : glycosyl transferase |
| PB : protein body | ✳ : folded lectin precursor |

tent lymphocyte mitogen and a leukoagglutinin with low erythrocyte activity, while $E_4$ is a strong erythroagglutinin but displays low mitogenic activity. The other isolectins $L_3E$, $L_2E_2$ and $LE_3$ display various erythroagglutinating and mitogenic activities, which are proportional to the quantities of E and L present (Leavitt et al. 1977).

By using immobilized Con A, Carter and Etzler (1975a) succeeded in fractionating affinity purified *Dolichos biflorus* lectin into two electrophoretically distinguishable forms. The B-form, which represents 19% of the lectin sample, does not bind to the Con A column, whereas the A-form is retained on the column and can further be resolved in subfractions differing in their affinity toward Con A. The same authors further showed that both the A- and the B-form specifically agglutinate, and with similar titers, type-A human erythrocytes, they give similar precipitin curves with hog blood A+H substances and they show similar inhibition curves with methyl-α-N-acetyl-galactosamine and N-acetyl-galactosamine. Besides, they have very similar molecular weights and amino acid compositions. These results thus clearly indicate that the A- and B-forms differ from each other in their glycosylation chains. The tetrameric *Dolichos biflorus* lectin is composed of equal amounts of two types of subunits, designated I and II. These subunits were shown to have a very similar amino acid composition and react identically upon immunodiffusion using antisera made against either subunit (Carter and Etzler 1975a, 1975b). Moreover, they have identical $NH_2$-terminal amino acid sequences (Etzler et al. 1977). It was shown by Carter and Etzler (1975c) and by Roberts et al. (1982) that all peptides, obtained by digestion or cleavage of the subunits respectively with V-8 protease from *Staphylococcus aureus* and CNBr, are identical except the COOH-terminal peptides. Subunit I has a slightly higher molecular weight than subunit II, and studies with carboxy-peptidases revealed that the subunits differ in their COOH-terminal amino acid residues (Carter and Etzler 1975c; Roberts et al. 1982). Despite the structural similarities between both subunits, only subunit I displays carbohydrate-binding activity (Etzler et al. 1981). These results made Etzler et al. (1981) suggest that both subunits may represent different degrees of completion, or modification of a common polypeptide chain. Furthermore, the difference in carbohydrate-binding capacity of both subunits indicates that either the carbohydrate-binding site is located near the COOH-terminus, or that its conformation is greatly influenced by this part of the subunit.

Apart from genetically encoded subunit differences or glycosylation heterogeneity, post-translational proteolytic processing can cumulate as well in multiple lectin forms. Affinity purified pea lectin from immature seeds for instance can be resolved in at least four different fractions by isoelectric focusing. The isolectin with a pI of 5.2 was shown to be built up of intact precursor subunits (van Driessche et al. 1983). Two other pea isolectins having a pI of 5.8 and 7.0 respectively, are both tetramers built up of two light chains (α) and two heavy chains (β) (Entlicher and Kocourek 1975; van Driessche et al., in preparation). The difference in their pI is caused by charge heterogeneity of the α-chains as a result of proteolytic processing at the COOH-terminus of a common α-chain precursor (van Driessche et al., in preparation). A fourth isolectin (Entlicher and Kocourek 1975; van Driessche et al., in preparation) is a hybrid species formed by α-subunit exchange between the two isolectins just mentioned.

As was clearly shown for *Sophora japonica* by Timberlake et al. (1980), multiple banding pattern may also be due to the existence of mixtures of charge variants at equilibrium. These authors observed that extraction and elution of the protein bands exhibiting the extreme mobilities of the various electrophoretic forms result in the reappearance of all compounds during electrophoretic separation. *Sophora japonica* lectin exhibits three protein bands upon polyacrylamide gel electrophoresis at pH 8.9. Timberlake et al. (1980) demonstrated that these mixtures at equilibrium can be shifted toward one single band upon specifically binding saccharide to the lectin, which indicates that this binding stabilizes one single conformer.

104

## 3.3    The Physiological Function of Legume Lectins

Different physiological functions have been attributed to Leguminosae lectins. It has been suggested that these lectins might be a special class of reserve proteins, enzymes, fungistatic and/or bacteriostatic agents, integral membrane and/or cell wall components and recognition signals implicated in the establishment of the legume-*Rhizobium* symbiosis.

This paragraph will give a survey of literature data which either favor or disapprove of the hypothesis mentioned above. It will be clear however that, in spite of the tremendous amount of information which accumulated during the last decade, the final word about the role lectins play in plants in general, and in legumes in particular, still remains to be said. However, it can be hoped that, as time goes by, the joined efforts of hundreds of lectinologists will be sublimated in a clear picture explaining why nature selected and conserved those proteins which we call lectins.

### 3.3.1    Are Legume Lectins Involved in the Rhizobium-Legume Symbiosis?

Legume roots can be infected by gram-negative bacteria belonging to the genus *Rhizobium*. This infection can lead to the induction and development of $N_2$-fixing root nodules. This symbiosis is characterized by a high degree of host-specificity, i. e., *Rhizobium leguminosarum* will nodulate the roots of *Vicia*, *Pisum*, *Lathyrus* and *Lens* species but not those of *Trifolium*; *R. japonicum* nodulates *Glycine* but not *Vicia*, *Trifolium* or *Medicago* species. Host-specificity implies that specific recognition phenomena between the root (site of infection) and the bacteria are involved. Because of its ecological, economical and fundamental importance, from all proposed functions mentioned above, the role of lectins as "mediators in legume-*Rhizobium* interaction" has by far attracted most attention.

Based on the fact that the lectin derived from the seeds of *Phaseolus vulgaris* bound to a strain of *Rhizobium phaseoli*, as was evidenced by the ability of lectin-treated bacteria to agglutinate erythrocytes, Hamblin and Kent (1973) were the first to hypothesize that lectins might bind bacteria to the roots of Leguminosae. Bohlool and Schmidt (1974) observed that isothiocyanate-labeled soybean seed lectin specifically combined with 22 out of 25 soybean-nodulating *Rhizobium* strains but not with any of the 23 strains which do not nodulate soybeans. These authors suggested that legume lectins might be responsible for the attachment of *Rhizobium* to the root by interacting specifically with a distinctive polysaccharide on the surface of the appropriate *Rhizobium* cell. The statement of Bohlool and Schmidt, now generally known as the "lectin-recognition hypothesis" was the start of a new area of intensive research and although a lot of information accumulated in favor, many observations still remain to be explained before the hypothesis will be generally accepted.

At this point of the discussion, it should be kept in mind that, in the experimental system used by Bohlool and Schmidt, the fluorescent-labeled soybean lectin consisted of a crude ammonium-sulfate precipitate of seed proteins which was later shown to be at least for 85 % contaminated by extraneous proteins (Dazzo and Truchet 1983). Furthermore, it was not shown by Bohlool and Schmidt whether the binding of the FITC-labeled SBA preparation was specific, i. e., inhibitable by haptenic sugars. Last but not least, there was no evidence at all whether or not soybean roots contain lectin and, if so, whether it is identical to the seed lectin. As a matter of fact, at the time of its formulation, the lectin-recognition hypothesis was just an attractive working model requiring the search for firm arguments in order to be accepted.

The validity of the hypothesis implies that: (1) the lectin should be present at these localizations where the nodulating *Rhizobia* attach, i.e., the surface of developing root hairs; (2) exclusively *Rhizobia* capable of infecting the roots of a particular legume should bind the corresponding lectin.

Without any doubt, the most convincing evidence in favor of the lectin-recognition hypothesis stems from the detailed investigations of the *Trifolium-Rhizobium trifolii* symbiosis. Since this work has been thoroughly reviewed during the last years (Dazzo 1981; Dazzo and Truchet 1983; Dazzo and Hollingsworth 1984), only the most important aspects will be mentioned here. Trifoliin A, the lectin isolated from seeds and seedling roots of white clover, is a glycoprotein which specifically agglutinates *Rhizobium trifolii* (Dazzo et al. 1978). Binding to and agglutination of the clovers symbiont can be inhibited by 2-deoxy-D-glucose and 2-amino-2,6-dideoxy-glucose. Immunohistochemical localization studies revealed that the lectin accumulates on the surface of root hairs, especially at the growing root hair tip (Dazzo et al. 1978). Since the lectin can be released from the roots with the haptenic sugar 2-deoxy-D-glucose, Dazzo et al. (1978) concluded that the lectin is associated with the root cell walls through its carbohydrate-binding site. The lectin cross-bridging model proposed by Dazzo and co-workers (Dazzo and Hubbel 1975b; Dazzo and Brill 1979) states that trifoliin A on the root hair tips cross-bridges similar antigenic determinants on the root hair cell and *Rhizobium trifolii*. A revised version of this model has been described by Dazzo and Truchet (1983), taking into account the findings that *Rhizobium trifolii* synthesizes multiple receptors for trifoliin A. Indeed, it was shown that capsular polysaccharides (Dazzo and Brill 1979) as well as lipopolysaccharides (Hrabak et al. 1981) bind clover lectin.

Although in the specific case of *Trifolium-Rhizobium* symbiosis, all evidence available until now favors the lectin-recognition hypothesis, for other legume-Rhizobia systems the situation is less clear, often confusing and even contradictory. Soon after the publication of Bohlool's and Schmidt's paper, Dazzo and Hubbell (1975b) reported that FITC-labeled Con A bound strongly to various nodulating and non-nodulating strains of *Rhizobium* regardless of their respective host, indicating that interactions between legume and *Rhizobia* may not always account for the observed host-range specificity. Similarly, Chen and Philips (1976) concluded that simple attachment of *Rhizobium* to legume roots is not the basis of host-symbiont specificity. They found that *R. leguminosarum* does not only attach to the tips of pea root hairs but also to root hairs of *Canavalia ensiformis* D.c., *Lupinus polyphyllus* Lindl., *Trifolium pratense* L. and *Medicago sativa* L. which are not infected by this bacterium. Besides, no relationship was found between lectin-*Rhizobium* interactions and the capacity of the bacteria to infect the plants. The fact that some nodulating *Rhizobium* strains do not bind the seed lectin of their host could be explained by Bhuvaneswari et al. (1977, 1978). They found that the expression of lectin receptors by *Rhizobia* strongly depends on both culture age and growth medium. Indeed, the biochemically specific binding sites for SBA are transient rather than constitutive components of the *Rhizobium japonicum* cell surface (Bhuvaneswari et al. 1977). Most strains of *Rhizobium japonicum* have the highest percentage of SBA-binding cells and the greatest number of SBA-binding sites per cell in the early and mid-log phases of growth. Besides, *R. japonicum* strains which do not bind the lectin at any stage of growth when cultured in artificial media, do so when grown in association with roots of soybean seedlings (Bhuvaneswari et al. 1978).

Similarly, Dazzo et al. (1979) showed that the transient appearance of trifoliin A on *Rhizobium trifolii* influences the ability of the bacteria to attach to clover roots. In subsequent studies (Hrabak et al. 1981) it was shown that the growth-phase-dependent binding of trifoliin A to *Rhizobium trifolii* in both cultures is related to the appearance of a unique determinant (2-amino-2,6-dideoxy-glucose) in the lipopolysaccharide of the bacteria at the beginning of the stationary phase. By analyzing the composition of the capsular and extracellular polysaccharides of *Rhizobium japonicum*, Mort and Bauer (1980) observed that methylation of galactose residues of the polysaccharides results in the loss of the SBA-binding capacity of bacteria. In addition to changes in the polysaccharide composition, there is a reduction of about 50 % in the percentage of cells which are encapsulated as the culture matures from the early to the late log phase of the growth. Since only encapsulated cells are able to bind SBA, Mort and Bauer (1980) concluded that the combination of changes in the capsular composition and loss of encapsulation can account for the loss of binding capacity during the growth of cultures of *Rhizobium japonicum*. According to Hrabak et al. (1981) the profound influence of the growth phase as well as the culture conditions on the composition of the lectin-binding polysaccharides of *Rhizobium* might be a major cause for conflicting data among laboratories testing the lectin-recognition hypothesis. Although these results can only explain why some investigators did, while others did not, find a correlation between lectin-binding and nodulating potency, they do not explain however why some lectins bind to non-nodulating *Rhizobia*.

In all studies mentioned above, as in Bohlool's and Schmidt's original paper, seed lectin was used though there was no priory reason to believe that root and seed lectins are identical proteins. In order to answer this most important question, several investigators have isolated, characterized and localized the root lectins from various legumes. A main problem in these studies is the fact that it is not always clearly established whether the isolated lectins are indeed real root lectins, i. e., synthesized and/or secreted by the roots, or whether they just represent a pool of seed lectin which leaked out of the seeds during imbibition (Fountain et al. 1977; Causse and Rougé 1983) and subsequentially adsorbed to the emerging roots.

Kijne et al. (1980, 1983) isolated from pea-root slime a lectin which was shown to be identical to the seed isolectin-2. The origin of the lectin was however not clearly established.

Hosselet et al. (1983) could unequivocally show that the roots of *Pisum sativum* contain an endogenous lectin which was found to be identical to the seed lectin as evidenced by SDS-electrophoresis, immunodiffusion, UV-spectroscopy, hemagglutination and its inhibition by carbohydrates. Besides the presence of an endogenous root lectin, these authors also found that 1.5 % of the total hemagglutinating activity of 4-day-old roots could be removed from the root surface with high molarity buffers. Since the addition of glucose (a specific inhibitor of pea-root and seed lectin) to the washing buffer did not affect the amount of recovered root surface lectin, Hosselet et al. (1983) concluded that the root surface lectin was not bound to the cell wall by its carbohydrate-binding site.

The root of soybean contains an intracellular lectin which is identical to the seed lectin (Gade et al. 1981, 1983). Studies on the distribution of this lectin suggest that it is associated with the outer root surface and concentrated in those segments of the roots at which root hair and early secondary roots are observed. Since this is the region at which *Rhizobium* attaches and where nodulation is probably initiated, the authors concluded that the root lectin might indeed be implicated in the interaction of developing soybean with *R. japonicum*. Stacey et al. (1980) came to the same conclusion by showing that the binding of *Rhizobium japonicum* to the roots and root hairs of *Glycine soja* is specifically inhibited by α-D-galactose and N-acetyl-D-galactosamine. Besides, they found that the root lectin is localized on the root surface. Root surface lectin from soybean seedlings could be stripped off by galactose and was shown to reappear at the surface, indicating that the intracellular root lectin represents a stored pool of lectin molecules which are secreted to the root surface (Gade et al. 1983). Similar results have been reported for *Trifolium repens* root lectin, but, unlike the soybean root lectin, it is synthesized de novo, at least in 2-day-old roots (Sherwood et al. 1984).

The secretion of endogenous root lectin is strongly influenced by the presence of $NO_3^-$ in the medium. Both for *Trifolium repens* (Sherwood et al. 1984) and *Pisum sativum* (Diaz et al. 1984) it has been shown that excess $NO_3^-$, which is known to inhibit nodulation, affects the distribution and activity of secreted root lectin in such a way that its interaction with nodulating *Rhizobia* is impaired.

The results just described can be considered as arguments in favor of the lectin-recognition hypothesis. However, based on the fact that root lectin could only be detected in very young seedlings, Rougé and Labroue (1977) and Pueppke et al. (1978) argued that lectins are not implicated in host-symbiont interactions. Indeed, many plants are more susceptible to nodulation when inoculation with *Rhizobia* is performed several weeks after sawing (Pueppke et al. 1978). It is now well established however (Borrebaeck 1984; Borrebaeck and Matthiasson 1983) that roots of 5–6 week-old *Phaseolus vulgaris* plants contain lectin which is identical to the seed lectin, as evidenced by hemagglutination, inhibition of hemagglutination, immunodiffusion, isolectin composition, and subunit structure. Furthermore, Hosselet et al. (1985) showed that pea lectin, or a cross-reacting material, is present at the root surface during the whole life cycle of the plant. Although only minute quantities could be detected by a sensitive ELISA technique, it might well be that the lectin is present in very restricted areas of the root surface, resulting in a high local concentration which may be physiologically significant.

Although the results described above indicate that in legumes such as *Pisum sativum, Glycine max, Phaseolus vulgaris* and *Trifolium repens*, the seed and root lectins are either identical or at least similar, especially with respect to their carbohydrate specificity, care must be taken for premature overgeneralizations. Indeed, Law and Strijdom (1984a, 1984b) showed that the seed and root lectins of *Lotononis bainessi* are clearly different molecules, with respect to their molecular weight, carbohydrate specificity, hemagglutinating and immunological properties, and binding capacity toward *Rhizobium*. While FITC-labeled root lectin bound to *Lotononis*-nodulating *Rhizobium* strains, but also to both a strain of *Rhizobium leguminosarum* and *Rhizobium japonicum*, the seed lectin failed to do so. Besides, the root lectin was shown to be localized at the tips of developing root hairs, lateral growth points of more mature root hairs and at the damaged edges of severed root hairs.

## 3.3.2   Are Legume Seed Lectins Storage Proteins?

It is common knowledge that seeds are an excellent source for the isolation of legume lectins. However, the reason why so high quantities of lectin are stored within the seed remains a question which has not been clearly answered until now, although several lines of evidence indicate that legume seed lectins may by used by germinating seeds as a nitrogen source. Subcellular fractionation of either dormant or inbibed seeds revealed that seed lectins of *Phaseolus vulgaris* (Bollini and Chrispeels 1978; Pusztai et al. 1979), *Vicia faba* and *Pisum sativum* (Weber et al. 1978, 1981; Weber and Neumann 1980) are localized in the protein bodies of the cotyledons. Although cellular fractionation gives valuable information on the subcellular localization of proteins, rupture of organelles during homogenization can hardly be avoided. Especially in the case of lectins, escaped molecules could be redistributed and bind to glycoconjugates of membranes, cell-wall fractions or, as was shown for *Ricinus communis* agglutinin, even penetrate into other organelles (Bowles et al. 1976; Köhle and Kauss 1979).

However, immunocytochemical localization studies at the ultrastructural level have confirmed that the seed lectins are mainly confined to the protein bodies. Using the immunogold technique, it was shown by Horisberger and Vonlanthen (1980) that soybean lectin was uniformly distributed in most of the protein bodies of the cotyledons and embryo axis. Similarly, in mature pea seeds, using the unlabeled peroxidase-antiperoxidase procedure, van Driessche et al. (1981) showed that the soluble pea seed lectin is localized in the protein bodies of both the parenchyma cells of the cotyledons and embryo axis. In immature pea seeds, however, at stages of seed development where the loading of protein bodies had not started yet, Smets et al. (1985), by using the immunogold method, localized pea lectin or a cross-reacting material in association with the cell walls of the embryo (Fig. 3.7). During later stages of seed development, the gold labels were detected in the endoplasmatic reticulum (see Fig. 3.7) and at the periphery of the protein bodies of both embryo and cotyledons (Smets et al. 1985) (see Fig. 3.7). Based on these studies, it was proposed (Smets et al. 1985) that the lectin can fulfill different development-dependent functions. The lectin which accumulates in the protein bodies during the later stages of seed maturation might function as reserve protein, whereas the lectin which is synthesized in the very early stages of development and which is found in the cell walls might represent receptors for glycosylated molecules, or enzymes participating in cell-wall metabolism. The lectin molecules localized at the rough endoplasmatic reticulum, especially in parenchyma cells of the cotyledons in immature seeds, should not be considered as structural components of this compartment but rather as molecules which are on their way to be deposited in the protein bodies. This view is indeed in accordance with our current knowledge on the synthesis and transport of lectins and storage proteins as reviewed by Chrispeels (1984).

A combination of indirect immunofluorescence and electron microscopy revealed that both the E- and L-type lectins of *Phaseolus vulgaris* are present in the parenchyma cells of the cotyledons inside the protein bodies. The matrix of all protein bodies contained both types of lectins. On the contrary, in vascular and axis cells the two types of lectins were mainly localized outside the protein bodies in the cytoplasm (Manen and Pusztai 1982). These authors suggested that the lectin present in the cotyledons might be a specific storage protein, while the vascular and axis lectins might play a more active metabolic function.

More recently however Greenwood et al. (1984) reported that PHA is exclusively localized in small vacuoles or protein bodies in all cortical parenchyma cells, vascular-bundle cells, as well as in epidermal cells of the embryonic axis while no lectin was found in the cytosol. In mature seeds, Con A too is present in the protein bodies of the storage parenchyma cells of the cotyledons but is absent from the embryonic tissues. During seed development, the lectin could also be localized on the endoplasmatic reticulum and Golgi apparatus (Herman and Shannon 1984).

By a combination of several techniques, such as subcellular fractionation, immunofluorescence and immunocytochemical techniques Etzler et al. (1984) were able to show that the seed lectin of *Dolichos biflorus* is mainly localized at the periphery of the protein bodies, and some lectin was found in association with the starch granules. These authors clearly demonstrated that lectin found in the cytoplasm is dependent upon the degree of protein body damage.

Since protein bodies are the site at which lectins are deposited in the seeds, it is not unreasonable to suppose that they are a special class of reserve proteins and/or that lectins are implicated in embryogenesis.

Howard et al. (1972), in studying the appearance and location of lentil lectin during the life cycle of the plant, suggested that the lectin is involved in maturation and germination processes. This suggestion was based on the fact that the lectin activity as measured by hemagglutination and immunodiffusion is mainly associated with the cotyledons and embryo. Upon germination, most of the remaining activity was detected in the cotyledons, whereas the stems and roots contained only small amounts of LcH. As the roots and stems developed and the cotyledons diminished in size, less LcH was associated with the whole seedling and at the stage of development where the cotyledons were no longer visible, no LcH could be detected in any part of the

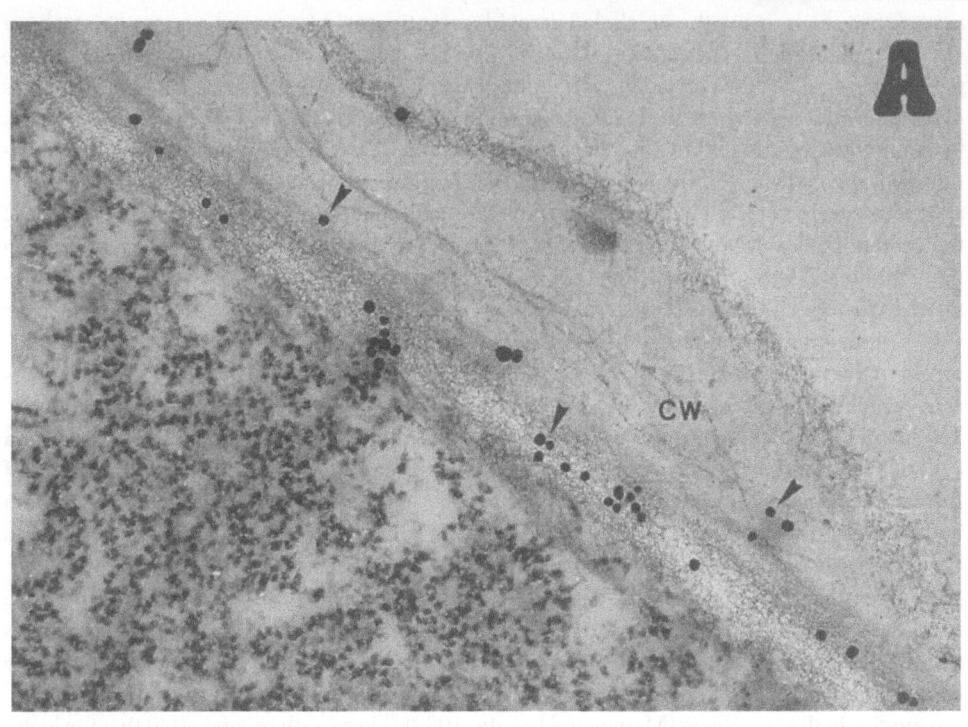

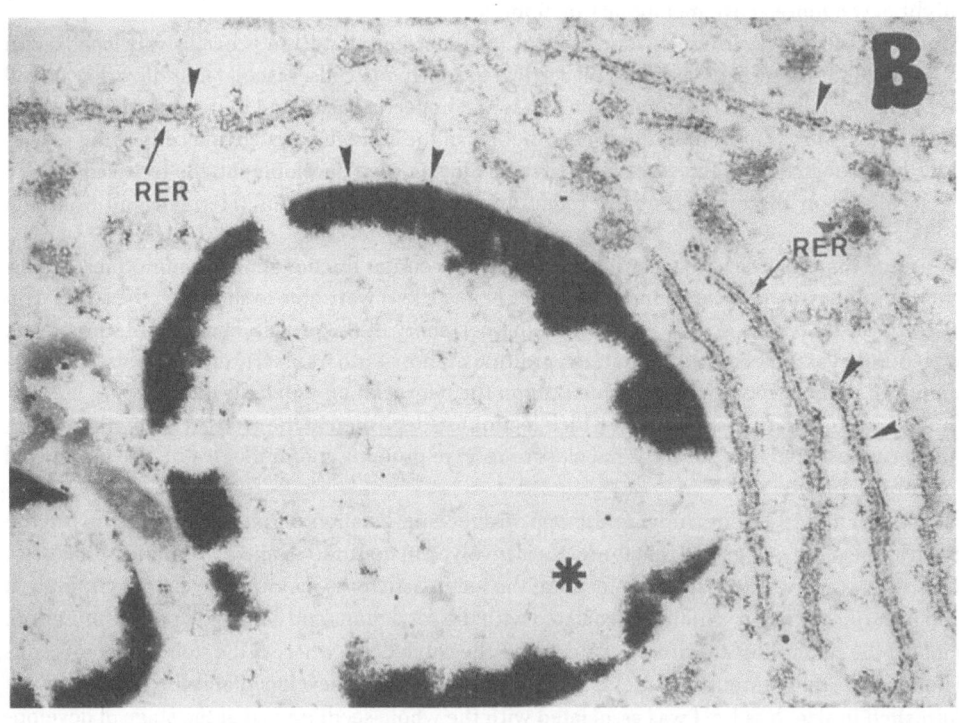

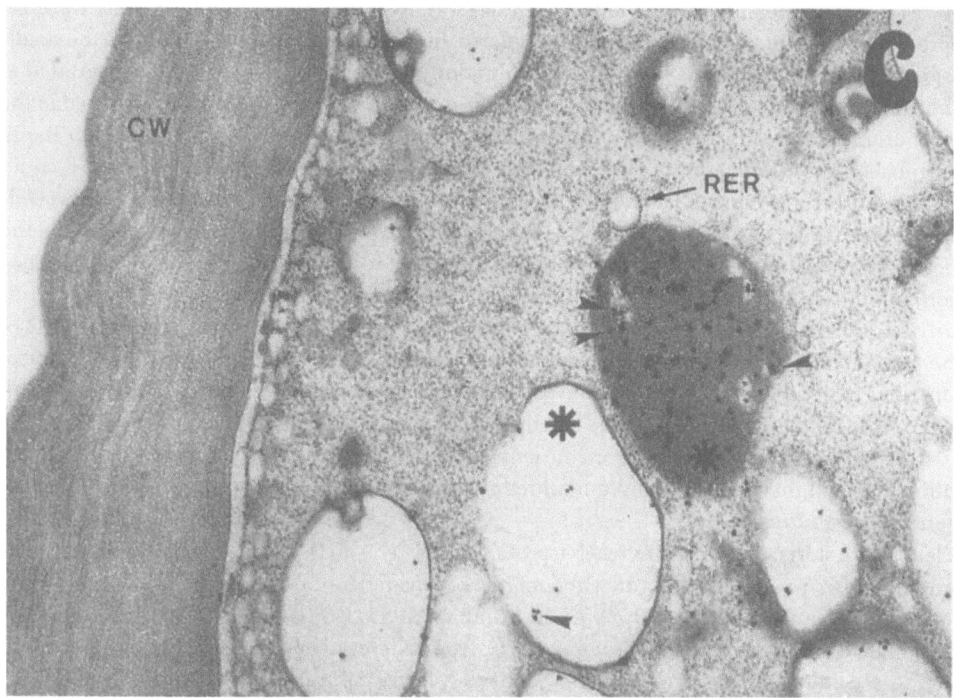

*Fig. 3.7* Localization of Pisum sativum lectin by immunogold staining

A) Pea lectin localization in embryo cells of young, immature seeds. Only the cell walls are labeled with gold granules; no intracellular labeling was noticed. Magnification: x 100,000.

B) Localization of pea lectin in cotyledons from immature seeds synthesizing lectin. Both protein bodies and rough endoplasmatic reticulum are labeled. Magnification: x 35,000.

C) Pea lectin localization in embryos from mature (24 hours imbibed) pea seeds. Protein bodies are labeled; cell walls and rough endoplasmatic reticulum are negative. Magnification: x 35,000.

Tissue was fixed in 0.2 M cacodylate buffer, pH 7.2, containing 2.5 % glutaraldehyde. For experimental details, see van Driessche et al. (1981) and Smets et al. (1985).

Abbreviations used:
CW : cell wall                    ◄: gold granules
RER: rough endoplasmatic reticulum    ✳ : protein bodies

plant. The lectin activity reappeared only in larger ripening seeds and fully ripe seeds while the other parts of the plant remained devoid of lectin. The results reported by Rougé and Plantefol (1974a, 1974b) pointed in the same direction in that lentil lectin disappears from roots, stems and leaves within 1 week after germination. These authors also concluded that seed lectins might be storage proteins and/or proteins implicated in embryogenesis. Besides, since the lectin so quickly disappeared from the roots, its involvement in host-symbiont recognition was rejected (Rougé and Labroue 1977).

The early studies on the distribution of lectins during the life cycle of the plant were conducted using hemagglutination and immunodiffusion. These methods are relatively insensitive compared with immunological binding assays, such as radio- and enzyme immunoassays. In spite of these sensitive detection methods, Pueppke et al. (1978) could only detect soybean lectin in all tissues of soybean seedlings until 2–3 weeks after germination. Similar results were reported by Causse et al. (1986). In *Dolichos biflorus*, high lectin levels are also found in the seeds, especially in the cotyledons, and upon germination the seed lectin appears to be degraded at about the same rate as the other reserve substances. Moreover no lectin could be detected in the roots of this plant at any stage of development (Talbot and Etzler 1978a). On the other hand, Borrebaeck and Mattiasson (1983), using a sensitive solid phase enzyme immunoassay, could find significant amounts of lectin in roots and stems of 3–10 week-old *Phaseolus vulgaris* seedlings, while the lectin content of leaves was low. Later it was shown (Borrebaeck 1984) that the lectins from vegetative tissues were identical to the seed lectin as evidenced by hemagglutination, inhibition of hemagglutination, immunodiffusion, isoelectric focusing and PAGE. Similarly, Hosselet et al. (1985) showed that pea lectin is present in all organs throughout the life of the plant. These results might indicate that the same lectin may play different physiological functions depending on its localization in the plant as was suggested previously (Smets et al. 1985). It is not clear at the moment whether the lectin in vegetative tissues is synthesized de novo or whether it is transported from the cotyledons and/or embryo axis. In the latter case, a dilution effect might explain the low quantities of lectin found throughout the life cycle of, for instance, *Pisum sativum*.

If lectins are stored in the seed as reserve proteins, then why do they have specific carbohydrate-binding sites? An answer to this question might be provided by the exciting work of Rüdiger and co-workers (Gansera et al. 1979; Shurz and Rüdiger 1985; Einhoff et al. 1986a, 1986b). They found that a discrete number of proteins, either from whole cotyledons or from isolated protein bodies, are bound to the corresponding Sepharose-lectin columns. These proteins, called lectin-binders, can be released from the affinity columns by raising the ionic strength of the washing buffer and, in the case of mannose/glucose-specific legume lectins, a second group of lectin-binders is desorbed by glucose or mannose. It was shown (Einhoff et al. 1986a, 1986b) that the lectin-binders belong either to the legumin type or vicilin type of storage proteins, or to both. Most remarkably, only a well defined subfraction of the storage proteins is retained on the columns. Besides storage protein-type lectin-binders, some glycosidases also behave as lectin-binders. It was proposed (Einhoff et al. 1986a, 1986b) that legume seed lectins serve as a packaging aid for storage proteins. Lectins might act as association centers around which the lectin-binders may first accumulate in an orderly manner. Since storage proteins strongly tend to undergo self-association, further packing could occur spontaneously. According to this model, even very low amounts of seed lectin might enable the packaging of reserve proteins. During germination it is supposed that, as a consequence of a rise in the ionic strength, the complexes (lectin + lectin-binder + storage proteins) are dissociated. Concomitantly, the storage proteins and lectins will be digested by proteolytic enzymes which are known to be present in the protein bodies as well (van der Wilden et al. 1980), while, by the action of glycosidases, the glycosylated reserve proteins in turn might become more susceptible to proteolytic degradation. This model elegantly explains why mannose/glucose-specific legume lectins contain a carbohydrate-binding site, i.e., it provides the means of specifically attaching the lectin to glycosylated lectin-binders. Lectins which are not mannose/glucose-specific do not bind the lectin-binders in a hapten-inhibitable way. In this class of lectins, the carbohydrate-binding capacity would be necessary to attach the lectin to the inner protein body membrane. Such a binding might of course also be operative in the case of the mannose/glucose-specific legume lectins, assuming that at least one carbohydrate-binding site remains free to interact with the glycosylated lectin-binders.

### 3.3.3 Are Legume Lectins Enzymes?

Hankins and Shannon (1978) were the first to report the isolation of a lectin from mung bean seeds that displayed strong enzymatic (α-galactosidase) activity. This lectin is a tetrameric glycoprotein composed of identical or nearly identical subunits. Since both the lectin and enzymatic activity copurified it was concluded that both activities reside on the same protein molecule. Indeed, the purification methods used (a combination of CM-cellulose chromatography and gel filtration on Sephacryl S-200) would most probably have resolved proteins differing in charge and molecular weight. Furthermore, upon SDS-electrophoresis only one single subunit species could be evidenced. The authors further excluded the possibility of trace contamination by showing that both the enzyme and the lectin had specific activities comparable to those of highly purified α-galactosidases and lectins from other sources. Finally, the two activities posses the same carbohydrate specificity, bind to erythrocytes in a hapten-inhibitable way, are equally heat-labile in the presence or in the absence of sugars, and are equally sensitive to the sulfydryl reagent p-chloro-mercurybenzoate. α-Galactosidase-hemagglutinins with properties similar to those of mung bean have also been isolated from other legume species (Hankins et al. 1980a).

It was found that several well-characterized legume lectins immunologically cross-react with the mung bean α-galactosidase-hemagglutinin, indicating that this lectin is closely related to at least some of the major seed lectins (Hankins et al. 1979).

Since structural relatedness may be indicative of a functional relationship, it was suggested that most or all legume seed lectins are in fact a homologous group of enzymes. Why then is enzymatic activity only demonstrated in some of them? According to Hankins and Shannon (1978) there might be different reasons. First of all, lectins might be enzymes that lost their enzymatic activity during the purification, but retained their substrate-binding capacity. Secondly, lectins might be enzymes with a very high degree of substrate specificity and, consequently, simple sugars, oligosaccharides, cellular receptors or glycosides might act as nonfunctional substrate analogs. Thirdly, since most of the time legume lectins are isolated from dry seeds, it might be possible that lectins are enzymatically inactive and acquire catalytic properties upon germination.

In order to get a more general view on the presence of enzyme-lectins, Hankins et al. (1980b) screened 20 legume species for α-galactosidase and hemagglutinin activity. These studies revealed that all species examined contained α-galactosidase which is immunologically related to the mung bean α-galactosidase-hemagglutinin. Furthermore, these α-galactosidases display very similar physical and kinetic properties to those of the mung bean enzyme lectin. However, only 13 out of 20 species examined contained hemagglutinating activity, and none of them possessed lectins with properties comparable to the mung bean α-galactosidase-hemagglutinin. Furthermore, del Campillo and Shannon (1982) could isolate an α-galactosidase-hemagglutinin from soybean seeds that was shown to be unrelated to the N-acetyl-D-galactosamine-specific soybean lectin. It may thus be supposed that the legume α-galactosidases, being widely distributed and strongly conserved with respect to physical and enzymatic characteristics, play an important role in legume seeds. On the other hand, the hemagglutinating activity, displayed by some of these enzymes, clearly is an in vitro property of the tetrameric α-galactosidase-I enzymes and is of no general physiological significance to the plant. It was shown by del Campillo et al. (1981) that the tetrameric mung bean α-galactosidase which displays both lectin and enzyme activity can be dissociated into monomers which retain their enzymatic activity but are devoid of hemagglutinating activity, indicating that the enzyme-lectin is an aggregated form of a monomeric α-galactosidase. In this context, the clot dissolving power of the mung bean α-

113

galactosidase lectin (Hankins and Shannon 1978) can be easily explained as being the result of the enzymatic modification of the erythrocyte receptors and a concomitant release of the lectin from the cells.

Unlike the mung bean type of α-galactosidase lectins, which may sometimes display lectin activity due to the presence of multiple catalytic sites, both the monomeric and tetrameric forms of the α-galactosidases of *Vicia faba* display true lectin activity (Dey et al. 1982a, 1982b; Dey and Pridham 1986). Whereas D-glucose and D-mannose and their derivatives produced no significant inhibition of the catalytic activity of the enzymes, they were shown to be potent inhibitors of the hemagglutination. Furthermore, all forms of *Vicia faba* α-galactosidases, i.e., forms I, $II^1$, $II^2$, are able to precipitate yeast mannan, glycogen and soluble starch, and enzyme I was found to be able to bind to starch granules leaving the catalytic site available for enzyme activity (Dey and Pridham 1986). These results clearly indicate that the lectin and the enzyme activities reside on different sites of the molecule. Equilibrium dialysis experiments have revealed that 8, 4 and 2 D-mannose molecules respectively are bound per molecule of enzyme form I, $II^1$ and $II^2$, explaining why both the tetrameric form (I) and the monomeric forms ($II^1$ and $II^2$) of the enzyme behave as hemagglutinins (Dey and Pridham 1986). It was proposed by Dey and Pridham (1986) that the lectin sites of the enzyme might be important for the enzyme in different ways. They might bind the enzyme to cell components and, as such, possibly regulate the activity of the catalytic sites, or they might be important in cross-linking the monomers and, in this way, modulate the enzyme activity and contribute to the quaternary structure of the enzyme.

### 3.3.4 Legume Lectins as Membrane and Cell-wall Components

It has been demonstrated by Bowles and Kauss (1975, 1976) that protein fractions extracted from membranes of mitochondria, Golgi complex, endoplasmatic reticulum and plasma membranes of growing mung bean hypocotyls, display hemagglutinin activity toward trypsinized rabbit erythrocytes. The hemagglutinating activity was shown to be most adequately released from the membranes by sonication in a mixture of EDTA/Triton-X 100. However, no sugars were found which caused complete inhibition of the agglutination, and no attempts were made to purify the lectins which were supposed to be integral membrane components. Similarly, lectins differing from the soluble seed lectins were reported to be present as membrane components throughout the life cycle of soybean and peanut (Bowles et al. 1979). The soluble lectins which were released from membranes by hapten inhibitors, high salt or low pH buffers, were effectively inhibited by monosaccharides, whereas the lectins released from membranes by detergent action could only be inhibited by glycoproteins.

It has been shown by Tsivion and Sharon (1981) that hemagglutination can also be mediated by various amphiphatic lipids. This lipid-mediated hemagglutination can be inhibited by certain glycoproteins while simple sugars have no effect. According to these authors strong hemagglutinating activity in crude extracts from biological material, which is inhibited by glycoproteins but not by simple sugars, cannot be taken as a proof for the existence of lectin in these extracts. In view of these results, it is not completely clear whether the hemagglutinating activity displayed by detergent extractions from membrane fractions mentioned above is really due to the presence of lectins. Besides, it was reported (Pueppke et al. 1981) that Triton-X 100, when used in conditions as described in previous studies (Bowles and Kauss 1975, 1976; Bowles et al. 1976, 1979) strongly interferes with both the hemagglutinating and the Lowry protein assay.

Pueppke et al. (1981) could unequivocally demonstrate the presence of membrane lectins in cotyledons of soybean lines producing the soluble 120,000 MW lectin. These membrane lectins, which are solubilized by Nonidet P-40, were found to be identical, or at least very similar, to the soluble lectin. Since little or no exogenously supplied $^3$H-SBA adhered to the membranes during extraction, and $^3$H-SBA did not bind to purified membranes, it is most unlikely that the membrane SBA originates from cytoplasmic contamination. Soybean lines, reported to be genotypically "lele" and which lack the soluble seed lectin, did not contain membrane SBA.

In growing and nongrowing hypocotyls of mung bean, Kauss and Glaser (1974) could demonstrate lectin activity in association with membranes of organelles and the cell walls. These authors hypothesized that these lectins might be implicated in the transport of cell wall polysaccharides enabling them to find their right place in the cell wall. Since the binding potential of cell wall associated lectins strongly diminished toward acidic pH-values, it was suggested that the cell wall lectins might function during extension growth by acting as a "gluing" substance being able to form reversible interconnections between cell wall polysaccharides. Later studies from the same laboratory (Kauss and Bowles 1976; Haapss et al. 1981) unequivocally revealed that, in spite of using five different isolation techniques, the hemagglutinating activity co-purified with that of α-galactosidase activity. From a comparison of the physicochemical properties of the hypocotyl cell wall and seed lectin of mung bean, it was concluded that both are very similar (Haapss et al. 1981). Besides, Haaps et al. (1981) found that the haptenic sugar galactose does not enhance the elution of the lectin from the cell wall, and neither prevents its binding to it. These findings suggest that the lectin is not located in the cell walls in vivo, but becomes bound to it during tissue homogenization.

The stems and leaves of *Dolichos biflorus* contain a lectin that cross-reacts with antibodies to the seed lectin ("cross-reacting material": CRM), and which is structurally similar to it (Etzler et al. 1977; Talbot and Etzler 1978b; Etzler and Borrebaeck 1980). However, despite the structural similarities, unlike the seed lectin, the CRM is not present in the developing seeds and only minute amounts are present in the dried seed (Roberts and Etzler 1984). Upon germination, the CRM accumulates at the apex of etiolated and green seedlings, in the epicotyl and the leaves. Minor amounts of CRM are also found in the cotyledons and the hypocotyl, but no CRM could be detected in the roots. During cell elongation, the amount of CRM increases in the first and the second internodes of the stems, while after the completion of elongation the CRM-content decreases. In 19-day old seedlings, most ot the CRM was found to be associated with elongating tissues (Roberts and Etzler 1984). Subcellular localization studies revealed that a significant portion of the stem and leaf lectins is found in association with the cell walls (Etzler et al. 1984), from which it can be extracted with NaCl-, detergent-, or EDTA-containing buffers. Some of the CRM was found at cytoplasmic sites. From these findings Etzler et al. (1984) suggested that the cell wall associated lectin might play a role in extension growth, while cytoplasmic CRM might be involved in the transport and assimilation of cell wall polysaccharides.

### 3.3.5   Are Legume Lectins Implicated in the Host's Defence Mechanism?

It was noticed by Sing and Schroth (1977) that the saprophytic bacterium *Pseudomonas putida* is first immobilized onto the cell walls and subsequently encapsulated in the intercellular spaces of *Phaseolus vulgaris* leaves by fibrillar structures originating from the cell walls. The phytopathogenic bacteria *Pseudomonas phaseolica* and *Pseudomonas tomato* on the other hand did not adhere to the cell walls, were not encapsulated and, unlike the saprophytic bacteria, devided in the intercellular spaces. These authors concluded that attachment and encapsulation may be a major defence mechanism in plants against bacteria. The fact that *Pseudomonas tomato*, which is a pathogen of tomatoes but not of beans, failed to be attached and encapsulated indicates that other defence mechanisms are also operative. However, since saprophytes are immobilized whereas pathogens are not, it is clear that specific recognition phenomena are involved.

Although lipopolysaccharides isolated from both saprophytic and pathogenic bacteria formed precipitation lines with *Phaseolus vulgaris* lectin, only the saprophyte was agglutinated. It was proposed by Sing and Schroth (1977) that this selective agglutinability is due to the presence of an extracellular polysaccharide in pathogens preventing the binding of the lectin to the bacterial surface. In view of the result reported by Borrebaeck (1984) that leaves of *Phaseolus vulgaris* contain lectin which is identical to the seed lectin, it is tempting to speculate that PHA might be implicated in the attachment of saprophytic bacteria to the cell walls adjacent to the intercellular spaces. However, in order to be accepted, this proposal put forward by Sing and Schroth (1977) must await detailed localization studies of PHA in the leaves.

From the seeds of soybean, cultivar Clark, Fett and Sequeira (1980a) succeeded in isolating a lectin that agglutinates *Xanthomonas phaseoli* var. *soyensis*, the causal agent of bacterial pustule disease of soybean. This lectin is clearly different from the classical $SBA_{120}$ in that it is not a hemagglutinin and furthermore displays different physicochemical properties such as solubility in ammonium sulfate, molecular weight, hapten specificity and immunological determinants.

In a subsequent brilliant, methodologically most important paper, Fett and Sequeira (1980b) conclusively showed that the bacterium agglutinating soybean seed lectin is not implicated in determining pathogen specificity, although their in vitro results might have pointed into this direction. These authors found for instance that crude fractions from Clark seed extracts strongly agglutinated most virulent strains of *Xanthomonas phaseoli* var. *soyensis*, whereas avirulent strains, or strains of *Pseudomonas glycinea* differing in pathogenicity from soybean, were not. Bacterial agglutination was inhibited by both purified lipopolysaccharide and extracellular polysaccharide from five *Xanthomonas phaseoli* var. *soyensis*, but not by these polysaccharides derived from other species of bacteria. However, upon screening a large number of *Xanthomonas phaseoli* var. *soyensis* strains, no correlation was found between virulence and agglutination by highly purified Clark seed lectin. Furthermore, the lectin could not be detected in the leaves of Clark soybean. Since the leaves are the site of bacterial invasion, Fett and Sequeira concluded that the lectin is not implicated in determining pathogen specificity.

At this point of the discussion, it is appropriate to remember that the lectin recognition hypothesis was initially based on in vitro interactions between legume seed lectins and *Rhizobia* without taking the localization and physicochemical properties of the lectin used into consideration. The results of Fett and Sequeira (1980b) thus clearly demonstrate the urgent need of combining in vitro studies with histological studies in order to get information on physiologically significant interactions. Indeed, lectins must be present in an active form at those locations where they are supposed to play their role and, in this special case, in concentrations which can cause an at least static effect on further growth of the pathogen.

During imbibition, soybean seeds were shown to release lectin (Fountain et al. 1977; Hwang et al. 1978; Causse and Rougé 1983) and protease inhibitors in their surroundings in an active form, and it was supposed that these exudated proteins could function in protecting the seeds and young seedlings against soil pathogens. In a comparative study on the lectin content of seeds from soybean cultivars which are either susceptible or resistant to *Phytophthora megasperma* var. *soyae* (Gibson et al. 1982), it was noticed that the latter contain more lectin, al-

116

though the variance within each group does not indicate a strong correlation of increased $SBA_{120}$ content in resistant cultivars. Most interestingly, seed lectin in resistant cultivars is secreted sooner and at higher concentrations into the surroundings during germination than in susceptible cultivars. The released lectin bound to mycelial cell walls in a hapten-inhibitable way; at high concentrations it was found able to retard mycelial growth and to retard slightly the growth of germination zoospores. Besides, the imbibate of resistant cultivars could slightly inhibit mycelial growth while that of susceptible cultivars was ineffective. Further evidence for the involvement of lectin in the observed growth inhibition was the fact that the active agent could be removed from the imbibate on an affinity resin which binds SBA. Besides, the material eluted from the column by galactose was shown to inhibit growth.

More recently, Brambl and Gade (1985) observed that several legume seed lectins bound to the germ tubes of asexual spores of *Neurospora crassa*, *Aspergillus amstelodami* and *Botryodiplodia theobromae*, and caused growth disruption during germination of the spores. These authors supposed that the lectins tested disrupted fungal growth by interfering with normal cell wall deposition and assembly.

## 3.3.6   Are Legume Lectins Dispensable Proteins?

From the evidence described in the previous paragraphs, it can be concluded that legume lectins might display several functions, i. e. at the root surface they might be implicated in the establishment of legume-*Rhizobium* interactions, in the seed they may function as storage proteins or as packaging aids involved in the organization of protein bodies, and in some instances they may display enzymatic properties. In the vegetative parts of the plant, their function is even less clear but some evidence has been presented which might point to their involvement in cell wall extension growth. Since some legume lectins have been shown to inhibit fungal growth, it might well be that legume lectins are components of the defense mechanism against phytopathogens. Although today no general consensus has been reached as to the physiological function of legume lectins, it is clear that, if these proteins fulfill an essential function, they should be present in all cultivars of those species where they are supposed to play this role.

The seeds of 102 lines of *Glycine max* L. were screened by Pull et al. (1978) for the presence of the 120,000 MW soybean lectin. The presence of this 120,000 MW lectin was estimated both qualitatively and quantitatively by measuring the amount of lectin eluted from an affinity column, by determining the hemagglutinating activity of seed extracts and by observing the binding of isothiocyanate-labeled proteins in extracts of seeds to cells of eight strains of *Rhizobium*. Besides, the extracts were submitted to polyacrylamide gel electrophoresis and the presence or the absence of a band migrating coincidently with SBA was scored. From these studies, Pull et al. (1978) noticed a wide variation in lectin content among the lines tested. Moreover, they discovered five lines which were completely devoid of lectin. These lectin-less soybean lines were however successfully nodulated by several strains of *Rhizobium japonicum*, which made Pull et al. (1978) conclude that the 120,000 MW soybean seed lectin is not required for the initiation of the soybean-*Rhizobium* symbiosis. Using immunodiffusion and radioimmunoassay, Su et al. (1980) confirmed and extended the results of Pull and co-workers by showing that the lectin-less soybean lines are also devoid of materials immunologically related to SBA. Besides, Su et al. (1980) were unable to detect SBA in the noncotyledon vegetative tissues, including the roots, of 20 soybean lines. Similarly, Stahlhut et al. (1981) found that 49% of the 559 screened *Glycine soja* lines are devoid of lectin immunologically related to the lectin from *Glycine max*, as was shown by Ouchterlony double diffusion. The remarkable discrepancy in the percentage of lectin-less lines of *Glycine max* and its wild ancestor *Glycine soja* (Stahlhut et al. 1981) suggests that upon domestication of soybean there was a strong selection pressure for the presence of lectin in the seeds. Alternatively, soybeans were domesticated from a narrow germ plasm of *Glycine soja* that contained lectin. From the fact that *Glycine soja* lines which are devoid of lectin germinate normally, grow, flower and set seeds just like the soybean plants possessing lectin, Stahlhut et al. (1981) concluded that the seed lectin is not involved in cell division, growth or differentiation. Since all lectin-less lines are equally well nodulated, it was also concluded that the seed lectin is not essential in legume-*Rhizobium* recognition.

Upon reexamination of the lectin-negative soybean lines described by Pull et al. (1978), Tsien and co-workers (Tsien et al. 1983) found that these lines do contain soybean lectin, although in quantities which are 1,000 to 10,000 times lower that those found in a reference cultivar "Chippewa". Hemagglutinating activity could be demonstrated in crude extracts from individual seeds of each cultivar, as well as in ammonium sulfate-precipitable protein fractions from seeds derived from individual plants. The hemagglutinating activity could be inhibited by galactose, anti-SBA-IgG and by a lectin-binding polysaccharide of *Rhizobium japonicum*. Furthermore, the lectin isolated from these cultivars was shown to display molecular properties very similar to the lectin isolated from the reference cultivar "Chippewa".

The studies of Goldberg et al. (1983) have shown that soybean lines which are considered to be lectin-negative according to Pull et al. (1978) contain the genes encoding the lectin. However, the Le-lines have a 3.4 kilobase DNA segment inserted in the L1-gene, resulting in a reduced transcription of the mutant gene and, consequently, in a lectin-less phenotype. This gene insertion has the structural features of a transposable element (Vodkin et al. 1983).

Studying wild-growing populations of *Phaseolus aborigineus* in Northwest-Argentinia and Bolivia, Brücher et al. (1969) could show that, while the seeds of five populations examined were uniformly positive in hemagglutinating reactions, in four populations up to 77% of the seeds did not display any detectable lectin activity. From these observations it was concluded that the bean lectins do not fulfill any vital role for the plant.

As for soybean, the seeds of some *Phaseolus vulgaris* cultivars were reported to be devoid of lectin. One of these lectin-less cultivars was studied in great detail by Pusztai et al. (1981, 1982). These authors observed that the seeds of the presumed lectin-free cultivar "Pinto-III" contain a lectin which is different from the common *Phaseolus vulgaris* lectin. The Pinto-III-lectin is a dimeric protein establishing a high reactivity toward pronase-treated rat erythrocytes but, unlike the common phytohemagglutinin, it does not bind to fetuin-Sepharose or the thyroglobulin-Sepharose. Furthermore, the Pinto-III lectin displays only slight immunological cross-reactivity toward phytohemagglutinin. The results of Pusztai et al. (1981, 1982) conclusively show that Pinto-III does not synthesize the common seed lectin. However, the roots of Pinto-III seedlings, especially when nodulated with *Rhizobium phaseoli*, synthesize small quantities of a lectin which is very similar to the common tetrameric *Phaseolus vulgaris* seed lectin in size and in immunological properties, which indicates that in the seeds of Pinto-III the genes of phytohemagglutinin are not expressed. More recently however Horowitz (1985) reported that Pinto-III seeds do contain phytohemagglutinin, although at a 40-fold reduced level when compared with a reference cultivar "Contender". Although both the "Pinto-III" and "Contender" cultivars were shown to contain the same number of lectin genes (Horowitz 1985), the amount of lectin mRNA was reduced tenfold in the "Pinto" cultivar, suggesting that the strongly reduced seed phytohemagglutinin level is due, at least in part, to a reduction in lectin mRNA.

# 3.4　　Epilogue

From the previous paragraphs it is obvious that our knowledge on legume lectins has increased tremendously during the last 15 years. The availability of new and efficient isolation procedures (Rüdiger, chapter 2) has made it possible to isolate lectins, especially from the seeds of a great variety of legumes, in sufficient quantities to allow their detailed physicochemical characterization. These studies have stimulated investigations on the use of legume lectins as valuable and highly appreciated tools in biological, biochemical and biomedical research for the demonstration (Roth 1986), isolation and characterization of glycosylated biomolecules and cells (Lis and Sharon 1986).

Even some 10 years ago it was general practice to state that "virtually nothing is known about the physiological function of legume lectins". It is true that even today a lot of questions remain to be answered with respect to the physiological functions of lectins in general, and legume lectins in particular. Nevertheless important progress has been made during the last decade and it can be expected that the investigations referred to in this paper were just a prelude to the full understanding of the roles lectins play in the life of legume plants. It is now well established that legume lectins are not restricted to the seed but may be present at all stages of the plants' life cycle, though in low quantities, and sensitive techniques such as RIA or ELISA have been used to demonstrate their presence. Comparative studies on seed and vegetative tissue lectins have revealed that they may be either identical or quite different, indicating that the expression of lectin genes is under meticulous developmental regulation. The understanding of these regulatory mechanisms will be of paramount importance in explaining why some legume species synthesize lectin, while others, ore even varieties of the same species, do not at a given stage of development. Indeed, the discovery of "lectin-less" varieties has prompted some investigators to conclude that legume lectins have only marginal, if any, role to play and have thus to be considered as dispensable proteins. However, detection methods whose sensitivity reaches far beyond that of hemagglutination and direct demonstration of lectin genes have shown that the lectin-less cultivars in fact do contain lectin genes which, as was shown in soybean, may be blocked by transposable elements. However, more recently even these cultivars were shown to contain a lectin not screened for in previous studies. Besides, it should be remembered that, when speaking about lectins, we are mentioning a group of proteins which are only operationally defined. Indeed, lectins are defined as "Sugar-binding proteins or glycoproteins of non-immune origin, which can agglutinate cells and/or precipitate glycoconjugates" (Goldstein et al. 1980), and no allusion is made to the physiological functions of these proteins. This implies that we might only be looking to a subclass of lectins which are easily detected by hemagglutination, ignoring several other lectins in plant tissues which have escaped our attention until now as a result of using screening systems whose "lectin-receptors" can reasonably be supposed to be at best analogs of the endogenous lectin-receptors. From the studies on the physiological function of legume lectins mentioned in this chapter, it seems not unreasonable to suppose that legume lectins play different functions in different tissues and at different stages of development of the plant. In the seed, available evidence indicates that the lectin may be implicated in the packaging of storage materials just before or at the onset of dormancy, while during germination the bulk of the seed lectin is used by the seedling to support its growth. Besides, in view of the high quantities of lectin present in seeds, a fungistatic and/or bacteriostatic function may also be envisaged. At later stages of development, a lectin, identical, or similar to, or different from the seed lectin may be synthesized by the root cells and exported to mediate the establishment of *Rhizobium*-legume symbiosis and to inhibit growth of root parasites in the immediate vicinity of the root. In the stems and leaves, legume lectins may play a role in cell elongation or defence mechanism. When more information becomes available on the nature of endogenous lectin-receptors, it might even become clear that legume lectins behave as enzymes which might profoundly affect differentiation by modifying lectin-receptors.

It is clear from this paper that until now most of the research on legume lectins has been focused on their isolation, physicochemical characterization and physiological function. A minority of investigators have been working on the significance and implication of the presence of legume lectins in the diet of man and his live-stock. The results of these studies have recently been reviewed by Pusztai (1986) and Liener (1986), and clearly indicate that legume seed lectins may be, at least in part, responsible for the poor nutritive value of some legume seeds which are generally considered to be a rich protein source.

*Acknowledgements*

The author is pleased to thank Prof. L. Kanarek and Prof. R. Dejaegere for helpful suggestions and for critically reading the manuscript. He is much indebted to his wife, Dr. Sonia Beeckmans, for her constant support and cooperation, the meticulous typing and for drawing the figures. Dr. Gerda Smets is gladly credited with providing previously unpublished photographs on the localization of pea lectin.

## 3.5     References

Agrawal BBL, Goldstein IJ (1968) Protein-carbohydrate interaction. VII. Physical and chemical studies on concanavalin A, the hemagglutinin of the jack bean. Arch Biochem Biophys 124:218–229

Allen LW, Svenson RH, Yachin S (1969) Purification of mitogenic proteins derived from *Phaseolus vulgaris*: isolation of potent and weak phytohemagglutinins possessing mitogenic activity. Proc Natl Acad Sci USA 63:334–341

Allen AK, Desai NN, Neuberger A (1976) The purification of the glycoprotein from the broad bean *(Vicia faba)* and a comparison of its properties with lectins of similar specificity. Biochem J 155:127–135

Baues RJ, Gray GR (1977) Lectin purification on affinity columns containing reductively aminated disaccharides. J Biol Chem 252:57–60

Baumann C; Rüdiger H, Strosberg AD (1979) A comparison of the two lectins from *Vicia cracca*. FEBS Lett 102:216–218

Baumann CM, Strosberg AD, Rüdiger H (1982) Purification and characterization of a mannose/glucose-specific lectin from *Vicia cracca*. Eur J Biochem 122:105–110

Bausch JN, Poretz RD (1977) Purification and properties of the hemagglutinin from *Maclura pomifera* seeds. Biochemistry 16:5790–5794

Becker JW, Reeke GN, Edelman GM (1971) Location of the saccharide-binding site of concanavalin A. J Biol Chem 246:6123–6125

Becker JW, Reeke GN, Wang JL, Cunningham BA, Edelman GM (1975) The covalent and three-dimensional structure of concanavalin A. III. Structure of the monomer and its interactions with metals and saccharides. J Biol Chem 250:1513–1524

Becker JW, Reeke GN, Cunningham BA, Edelman GA (1976) New evidence on the location of the saccharide-binding site of concanavalin A. Nature (London) 259:406–409

Bhattacharyya L, Brewer CF (1984) Preparation and characterization of $Ca^{2+}$-$Zn^{2+}$-derivatives of lentil and pea lectins and comparison with the native forms. Biochem Biophys Res Commun 124:857–862

Bhattacharyya L, Brewer CF (1985) Preparation and properties of metal-ion derivatives of the lentil and pea lectins. Biochemistry 24:4974–4980

Bhuvaneswari TV, Bauer WD (1978) Role of lectins in plant-microorganism interactions. III. Influence of rhizosphere/rhizoplane culture conditions on the soybean lectin-binding properties of Rhizobia. Plant Physiol 62:71–74

Bhuvaneswari TV, Pueppke SG, Bauer WD (1977) Role of lectins in plant-microorganism interactions. I. Binding of soybean lectin to Rhizobia. Plant Physiol 60:486–491

Blacik LJ, Breen M, Weinstein HG, Sittig RA, Cole M (1978) An anti-A$_1$ lectin in the seeds of *Amphicarpaea bracteata*. Biochim Biophys Acta 538:225–231

Blobel G, Dobberstein B (1975) Transfer of proteins across membranes. II. Reconstitution of functional rough microsomes from heterologous components. J Cell Biol 67:852–862

Bloch R, Jenkins J, Roth J, Burger MM (1976) Purification and characterization of two lectins from *Caragana arborescens* seeds. J Biol Chem 251:5929–5935

Blumberg S, Tal N (1976) Effect of divalent metal-ions on the digestibility of concanavalin A by endopeptidases. Biochim Biophys Acta 453:357–364

Blumberg S, Hildesheim J, Yariv J, Wilson KJ (1972) The use of 1-amino-L-fucose bound to Sepharose in the isolation of L-fucose-binding proteins. Biochim Biophys Acta 264:171–176

Bohlool BB, Schmidt EL (1974) Lectins: a possible basis for specificity in the *Rhizobium*-legume root nodule symbiosis. Science 185:269–271

Bollini R, Chrispeels MJ (1978) Characterization and subcellular localization of vicilin and phytohemagglutinin, the two major reserve proteins of *Phaseolus vulgaris* L. Planta 142:291–298

Borrebaeck CAK (1984) Detection and characterization of a lectin from non-seed tissue of *Phaseolus vulgaris*. Planta 161:223–228

Borrebaeck CAK, Mattiasson B (1983) Distribution of a lectin in tissues of *Phaseolus vulgaris*. Physiol Plantarum 58:29–32

Borrebaeck CAK, Lönnerdal B, Etzler ME (1981) Metal-ion content of *Dolichos biflorus* lectin and effect of divalent cations on lectin activity. Biochemistry 20:4119–4122

Bourillon R, Font J (1968) Purification and propriétés physicochimiques d'une phytohemagglutinine de *Robinia pseudoaccacia*. Isolement d'un glycopeptide actif. Biochim Biophys Acta 154:28–39

Bowles DJ, Kauss H (1975) Carbohydrate-binding proteins from cellular membranes of plant tissue. Plant Sci Lett 4:411–418

Bowles DJ, Kauss H (1976) Characterization, enzymatic and lectin properties of isolated membranes from *Phaseolus aureus*. Biochim Biophys Acta 443:360–374

Bowles DL, Schnarrenberger C, Kauss H (1976) Lectins as membrane components of mitochondria from *Ricinus communis*. Biochem J 160:375–382

Bowles DL, Lis H, Sharon N (1979) Distribution of lectins in membranes of soybean and peanut plants. I. General distribution in root, shoot and leaf tissue at different stages of growth. Planta 145:193–198

Brambl R, Gade W (1985) Plant seed lectins disrupt growth of germinating fungal spores. Physiol Plantarum 64:402–408

Brücher O, Wecksler M, Levy A, Palozzo A, Jaffé WG (1969) Comparison of phytohemagglutinins in wild beans *(Phaseolus aborigineus)* and in common beans *(Phaseolus vulgaris)* and their inheritance. Phytochemistry 8:1739–1743

Carrington DM, Auffret A, Hanke DE (1985) Polypeptide ligation occurs during post-translational modification of concanavalin A. Nature (London) 313:64–67

Carter WG, Etzler ME (1975a) Isolation, characterization and subunit structures of multiple forms of *Dolichos biflorus* lectin. J Biol Chem 250:2756–2762

Carter WG, Etzler ME (1975b) Isolation and characterization of subunits from the predominant form of *Dolichos biflorus* lectin. Biochemistry 14:2685–2689

Carter WG, Etzler ME (1975c) Isolation and characterization of cyanogen bromide fragments and a glycopeptide from the *Dolichos biflorus* lectin. Biochemistry 14:5118–5122

Catsimpoolas N, Meyer EW (1969) Isolation of soybean hemagglutinin and demonstration of multiple forms by isoelectric focusing. Arch Biochem Biophys 132:279–285

Causse H, Rougé P (1983) Lectin release from imbibed soybean seed and its possible function. In: Bøg-Hansen TC, Spengler GA (eds) Lectins: Biology, Biochemistry, Clinical Biochemistry, de Gruyter, Berlin (W), Vol 3, pp 559–572

Causse H, Lemoine A, Rougé P (1986) Quantification of soybean seed lectin in soybean tissues during the life-cycle of the plant, by an enzyme-linked immunosorbent assay (ELISA). In: Bøg-Hansen TC, van Driessche E (eds) Lectins: Biology, Biochemistry, Clinical Biochemistry, de Gruyter, Berlin (W), Vol 5, pp 667–675

Chen APT, Philips DA (1976) Attachment of Rhizobium to legume roots as the basis for specific interactions. Physiol Plantarum 38:83–88

Cheung G, Haratz A, Katar M, Skrokov R, Poretz RD (1979) Purification and properties of the hemagglutinin from *Wisteria floribunda* seeds. Biochemistry 18:1646–1650

Chou PY, Fasman GD (1974) Prediction of protein conformation. Biochemistry 13:224–245

Chrispeels MJ (1983) The Golgi apparatus mediates the transport of phytohemagglutinin to the protein bodies in bean cotyledons. Planta 158:140–151

Chrispeels MJ (1984) Biosynthesis, processing and transport of storage proteins and lectins in cotyledons of developing legume seeds Phil. Trans. R. Soc. Lond. B. 304:309–322

Cummings RD, Kornfeld S (1982a) Characterization of the structural determinants required for the high affinity interaction of asparagine-linked oligosaccharides with immobilized *Phaseolus vulgaris* leukoagglutinating and erythroagglutinating lectins. J Biol Chem 257:11230–11234

Cummings RD, Kornfeld S (1982b) Fractionation of asparagine-linked oligosaccharides by serial lectin-agarose affinity chromatography. A rapid, sensitive and specific technique. J Biol Chem 257:11235–11240

Cunningham BA, Wang JL, Pflumm MN, Edelman GM (1972) Isolation and proteolytic cleavage of the intact subunit of concanavalin A. Biochemistry 11:3233–3239

Cunningham BA, Wang JL, Waxdal MJ, Edelman GM (1975) The covalent and three-dimensional structure of concanavalin A. II. Amino acid sequence of cyanogen bromide fragment $F_3$. J Biol Chem 250:1503 to 1512

Cunningham BA, Hemperly JJ, Hopp TP, Edelman GM (1979) Favin versus concanavalin A: circularly permuted amino acid sequences. Proc Natl Acad Sci 76:3218–3222

Dazzo FB (1981) Bacterial attachment as related to cellular recognition in the *Rhizobium*-legume symbiosis. J Supramol Struct Cell Biochem 16:29–41

Dazzo FB, Brill W (1979) Bacterial polysaccharides which binds *Rhizobium trifolii* to clover root hairs. J Bacteriol 137:1362–1373

Dazzo FB, Hollingsworth RE (1984) Trifoliin A and carbohydrate receptors as mediators of cellular recognition in the *Rhizobium trifolii*-clover symbiosis. Biol Cell 51:267–274

Dazzo FB, Hubbell DH (1975a) Cross-reactive antigens and lectin as determinants of symbiotic specificity in the *Rhizobium*-clover association. Applied Microbiol 30:1017–1033

Dazzo FB, Hubbell DH (1975b) Concanavalin A: lack of correlation between binding to *Rhizobium* and specificity in the *Rhizobium*-legume symbiosis. Plant and Soil 43:713–717

Dazzo FB, Truchet GL (1983) Interactions of lectins and their saccharide receptors in the *Rhizobium*-legume symbiosis. J Membrane Biol 73:1–16

Dazzo FB, Yanke WE, Brill WJ (1978) Trifoliin: a *Rhizobium* recognition protein from white clover. Biochim Biophys Acta 539:276–286

Dazzo FB, Urbano M, Brill W (1979) Transient appearance of lectin receptors on *Rhizobium trifolii*. Curr Microbiol 2:15–20

Debray H, Decout D, Strecker G, Spik G, Montreuil J (1981) J.: Specificity of twelve lectins towards oligosaccharides and glycopeptides related to N-glycosylproteins. Eur J Biochem 117:41–55

Del Campillo E, Shannon LM (1982) An α-galactosidase with hemagglutinin properties from soybean seeds. Plant Physiol 69:628–631

Del Campillo E, Shannon LM, Hankins CN (1981) Molecular properties of the enzymic phytohemagglutinin of mung bean. J Biol Chem 256:7177–7180

Delmotte FM, Goldstein IJ (1980) Improved procedures for purification of the *Bandeiraea simplicifolia* I isolectins and *Bandeiraea simplicifolia* II lectin by affinity chromatography. Eur J Biochem 112:219–223

Dey PM, Pridham JB (1986) *Vicia faba* α-galactosidases with lectin activities: an overview. In: Bøg-Hansen TC, van Driessche E (eds) Lectins: Biology, Biochemistry, Clinical Biochemistry, de Gruyter, Berlin (W), Vol 5, pp 161–170

Dey PM, Naik S, Pridham JB (1982a) The lectin nature of α-galactosidases from *Vicia faba* seeds. FEBS Lett 150:233–237

Day PM, Pridham JB, Sumar N (1982b) Multiple forms of *Vicia faba* α-galactosidases and their relationships. Phytochemistry 21:2195–2199

Diaz CL, Lems-van-Kan P, van der Schaal IAM, Kijne JW (1984) J.W.: Determination of pea (*Pisum sativum* L.) root lectin using an enzyme-linked immunoassay. Planta 161:302–307

Dombrink-Kurtzmann MA, Dick WE, Burton KA, Cadmus MC, Slodki ME (1983) A soybean lectin having 4-0-methyl-D-glucuronic acid specificity. Biochem Biophys Res. Comm. 111:798–803

Duk M, Lisowska E (1981) *Vicia* graminea anti-N lectin: partial characterization of the purified lectin and its binding to erythrocytes. Eur J Biochem 118:131–136

Dupuis G, Leclair B (1982) Studies on Phaseolus vulgaris phytohemagglutinin. Structural requirements for simple sugars to inhibit the agglutination of human group A erythrocytes. FEBS Lett 144:29–32

Edelman GM, Wang JL (1978) Binding and functional properties of concanavalin A and its derivatives. III. Interactions with indoleacetic acid and other hydrophobic ligands. J Biol Chem 253:3016–3022

Edelman GM, Cunningham BA, Reeke GN, Becker JW, Waxdal MJ, Wang JL (1972) The covalent and three-dimensional structure of concanavalin A. Proc Natl Acad Sci USA 69:2580–2584

Einhoff W, Fleischmann G, Freier T, Kummer H, Rüdiger H (1986a) Interactions of leguminous seed lectins with seed proteins – Lectins as packing aids of storage proteins. In: Bøg-Hansen TC, van Driessche E (eds) Lectins: Biology, Biochemistry, Clinical Biochemistry, de Gruyter, Berlin (W), Vol 5, pp 45–52

Einhoff W, Fleischmann G, Freier T, Kummer H, Rüdiger H (1986b) Interactions between lectins and other components of leguminous protein bodies. Biol Chem Hoppe-Seyler 367:15–25

Entlicher G, Kocourek J (1975) Studies on phytohemagglutinins. XXIV. Isoelectric point and hybridization of the pea (*Pisum sativum* L.) isophytohemagglutinins. Biochim Biophys Acta 393:165–169

Entlicher G, Koštiř JV, Kocourek J (1970) Studies on phytohemagglutinins. III. Isolation and characterization of hemagglutinins from the pea (*Pisum sativum* L.). Biochim Biophys Acta 221:272–281

Ersson B (1977) A phytohemagglutinin from sunn hemp seeds *(Crotalaria juncea)*. II. Purification by a high capacity biospecific affinity adsorbent and its physicochemical properties. Biochim Biophys Acta 494:51 to 60

Ersson B, Asperg K, Porath J (1973) The phytohemagglutinin from sunn hemp seeds *(Crotalaria juncea)*. Purification by biospecific affinity chromatography. Biochim Biophys Acta 310:446–452

Etzler ME, Borrebaeck CAK (1980) Carbohydrate-binding activity of a lectin-like glycoprotein from stems and leaves of *Dolichos biflorus*. Biochem Biophys Res Comm 96:92–97

Etzler ME, Kabat EA (1970) Purification and characterization of a lectin (plant hemagglutinin) with blood group A specificity from *Dolichos biflorus*. Biochemistry 9:869–877

Etzler ME, Talbot CF, Ziaya PR (1977) NH$_2$-Terminal sequences of the subunits of *Dolichos biflorus* lectin. FEBS Lett 82:39–41

Etzler ME, Gupta S, Borrebaeck C (1981) Carbohydrate-binding properties of the *Dolichos biflorus* lectin and its subunits. J Biol Chem 256:2367–2370

Etzler ME, MacMillan S, Scates S, Gibson DM, James JR DW, Cole D, Thayer S (1984) Subcellular localisations of two *Dolichos biflorus* lectins. Plant Physiol 76:871–878

Falasca A, Franceschi C, Rossi CA, Stirpe F (1979) Purification and partial characterization of a mitogenic lectin from *Vicia sativa*. Biochim Biophys Acta 577:71–81

Felsted RL, Egorin MJ, Leavitt RD, Bachur NR (1977) Recombinations of subunits of *Phaseolus vulgaris* isolectins. J Biol Chem 252:2967–2971

Fett WF, Sequeira L (1980a) A new bacterial agglutinin from soybean. I. Isolation, partial purification and characterization. Plant Physiol 66:847–852

Fett WF, Sequeira L (1980b) A new bacterial agglutinin from soybean. II. Evidence against a role in determining pathogen specificity. Plant Physiol 66:853–858

Fish WW, Hamlin LM, Miller RL (1978) The macromolecular properties of peanut agglutinin. Arch Biochem Biophys 190:693–698

Fleischmann G, Rüdiger H (1986) Isolation, resolution and partial characterization of two *Robinia pseudoacacia* seed lectins. Biol Chem Hoppe-Seyler 367:27–32

Fliegerova O, Salvetova A, Ticha M, Kocourek J (1974) Studies on phytohemagglutinins. XIX. Subunit structure of the lentil isophytohemagglutinins. Biochim Biophys Acta 351:416–426

Foriers A, van Driessche E, de Neve R, Kanarek L, Strosberg AD, Wuilmart C (1977a) The subunit structure and N-terminal sequences of the α- and β-subunits of the lentil lectin *(Lens culinaris)*. FEBS Lett 75:237–240

Foriers A, Wuilmart C, Sharon N, Strosberg AD (1977b) Extensive sequence homologies among lectins from leguminous plants. Biochem Biophys Res Comm 75:980–986

Foriers A, de Neve R, Kanarek L, Strosberg AD (1978) Common ancestor for concanavalin A and lentil lectin? Proc Natl Acad Sci USA 75:1136–1139

Foriers A, Lebrun E, van Rapenbusch R, de Neve R, Strosberg AD (1981) The structure of the lentil *(Lens culinaris)* lectin. Amino acid sequence determination and prediction of the secondary structure. J Biol Chem 256:5550–5560

Fornstedt N, Porath J (1975) Characterization studies on a new lectin found in seeds of *Vicia ervilia*. FEBS Lett 57:187–191

Fountain DW, Foard DE, Replogle WD, Yang WK (1977) Lectin release by soybean seeds. Science 197:1185–1187

Frost RG, Reitherman RW, Miller AL, O'Brien JS (1975) Purification of *Ulex europaeus* hemagglutinin I by affinity chromatography. Anal Biochem 69:170–179

Gachelin G, Goldstein L, Hofnung D, Kalb AJ (1972) Implication of two histidines in transition-metal binding in concanavalin A. Eur J Biochem 30:155–162

Gade W, Jack MA, Dahl JB, Schmidt EL, Wold F (1981) The isolation and characterization of a root lectin from soybean *(Glycine max* L., cultivar Chippewa). J Biol Chem 256:12905–12910

Gade W, Schmidt EL, Wold F (1983) Evidence for the existence of an intracellular root-lectin in soybeans. Planta 158:108–110

Galbraith W, Goldstein IJ (1970) Phytohemagglutinins: a new class of metalloproteins. Isolation, purification, and some properties of the lectin from *Phaseolus lunatus*. FEBS Lett 9:197–201

Galbraith W, Goldstein IJ (1972) Phytohemagglutinin of the lima bean *(Phaseolus lunatus)*. Isolation, characterization and interaction with type A blood-group substance. Biochemistry 11:3976–3984

Gansera R, Schurz H, Rüdiger H (1979) Lectin-associated proteins from the seeds of Leguminosae. Hoppe-Seyler's Z Physiol Chem 360:1579–1585

Gebauer G, Schiltz E, Schimpl A, Rüdiger H (1979) Purification and characterization of a mitogenic lectin and a lectin-binding protein from *Vicia sativa*. Hoppe-Seyler's Z Physiol Chem 360:1727–1735

Gebauer G, Schiltz E, Rüdiger H (1981) The amino-acid sequence of the α-subunit of the mitogenic lectin from *Vicia sativa*. Eur J Biochem 113:319–325

Gibson DM, Stack S, Krell K, House J (1982) A comparison of soybean agglutinin in cultivars resistant and susceptible to *Phytophthora megasperma* var. sojae (race 1). Plant Physiol 70:560–566

Goldberg RB, Hoschek G, Vodkin LO (1983) An insertion sequence blocks the expression of a soybean lectin gene. Cell 33:465–475

Goldstein IJ, Hayes CE (1978) The lectins: carbohydrate-binding proteins of plants and animals. Adv Carbohydrate Chem Biochem 35:127–340

Goldstein IJ, Poretz RD (1986) Isolation, physicochemical characterization and carbohydrate-binding specificity of lectins. In: Liener E, Sharon N, Goldstein IJ (eds) The Lectins: Properties, Functions and Applications in Biology and Medicine, Academic Press, London, pp 33–247

Goldstein IJ, Hughes RC, Monsigny M, Osawa T, Sharon N (1980) What should be called a lectin? Nature (London) 285:66

Greenwood JS, Keller GA, Chrispeels MJ (1984) Localization of phytohemagglutinin in the embryonic axis of *Phaseolus vulgaris* with ultra-thin cryosections embedded in plastic after indirect immunolabeling. Planta 162:548–555

Grubhoffer L, Ticha M, Kocourek J (1981) Isolation and properties of a lectin from the seeds of hairy vetch *(Vicia villosa* Roth). Biochem J 195:623–626

Gunther GR, Wang JL, Yahara I, Cunningham BA, Edelman GM (1973) Concanavalin A derivatives with altered biological activities. Proc Natl Acad Sci USA 70:1012–1016

Gupta BKD, Chatterjee-Ghose R, Sen A (1980) Purification and properties of mitogenic lectins from seeds of *Lathyrus sativus* Linn. (chickling vetch). Arch Biochem Biophys 201:137–146

Haapss D, Frey R, Thiesen M, Kauss H (1981) Partial purification of hemagglutinin associated with cell walls from hypocotyls of *Vigna radiata*. Planta 151:490–496

Hamblin J, Kent SP (1973) Possible role of phytohemagglutinin in *Phaseolus vulgaris* L. Nature (London) 245:28–30

Hammarström S, Murphy LA, Goldstein IJ, Etzler Me (1977) Carbohydrate-binding specificity of four N-acetyl-D-galactosamine-"specific" lectins: *Helix pomatia* A hemagglutinin, soy bean agglutinin, lima bean lectin, and *Dolichos biflorus* lectin. Biochemistry 16:2750–2755

Hammarström S, Hammarström ML, Sundblad G, Arnarp J, Lönngren J (1982) Mitogenic leukoagglutinin from *Phaseolus vulgaris* binds to a pentasaccharide unit in N-acetyl-lactosamine-type glycoprotein glycans. Proc Natl Acad Sci USA 79:1611–1615

Hankins CN, Shannon LM (1978) The physical and enzymatic properties of a phytohemagglutinin from mung beans. J Biol Chem 253:7791–7797

Hankins CN, Kindinger JI, Shannon LM (1979) Legume lectins. I. Immunological cross-reactions between the enzymic lectin from mung beans and other well characterized legume lectins. Plant Physiol 64:104–107

Hankins CN, Kindinger JI, Shannon LM (1980a) Legume α-galactosidases which have hemagglutinin properties. Plant Physiol 65:618–622

Hankins CN, Kindinger JI, Shannon LM (1980b) Legume α-galactosidase forms devoid of hemagglutinating activity. Plant Physiol 66:375–378

Hapner KD, Robbins JE (1979) Isolation and properties of a lectin from sainfoin (Onobrychis viciifolia, Scop.). Biochim Biophys Acta 580:186–197

Hardman KD, Ainsworth CF (1972) Structure of concanavalin A at the 2.4 A resolution. Biochemistry 11:4910–4919

Hardman KD, Ainsworth CF (1973) Binding of nonpolar molecules by crystalline concanavalin A. Biochemistry 12:4442–4448

Hardman KD, Ainsworth CF (1976) Structure of the concanavalin A-methyl-α-D-mannopyranoside complex at 6-Å resolution. Biochemistry 15:1120–1128

Hassing GS, Goldstein IJ, Marini M (1971) The role of protein carboxyl groups in carbohydrate-concanavalin A interaction. Biochim Biophys Acta 243:90–97

Hayes CE, Goldstein IJ (1974) An α-D-galactosyl-binding lectin from *Bandeiraea simplicifolia* seeds. Isolation by affinity chromatography and characterization. J Biol Chem 249:1904–1914

Hemperly JJ, Hopp TP, Becker JW, Cunningham BA (1979) The chemical characterization of favin, a lectin isolated from *Vicia faba*. J Biol Chem 254:6803–6810

Hemperly JJ, Mostov KE, Cunningham BA (1982) In vitro translation and processing of a precursor form of favin, a lectin from *Vicia faba*. J Biol Chem 257:7903–7909

Herman EM, Shannon LM (1984) Immunocytochemical localization of concanavalin A in developing jack-bean cotyledons. Planta 161:97–104

Herman EM, Shannon LM, Chrispeels MJ (1985) Concanavalin A is synthesized as a glycoprotein precursor. Planta 165:23–29

Herrmann MS, Behnke WD (1980) Physical studies on three lectins from the seeds of *Abrus precatorius*. Biochim Biophys Acta 621:43–52

Herrmann MS, Richardson CE, Setzler LM, Behnke WD, Thompson RE (1978) A circular dichroic investigation of the secondary structure of lectins. Biopolymers 17:2107–2120

Higgins TJV, Chandler PM, Zurawski G, Button SC, Spencer D (1983a) The biosynthesis and primary structure of pea seed lectin. J Biol Chem 258:9544–9549

Higgins TJV, Chrispeels MJ, Chandler PM, Spencer D (1983b) Intracellular sites of synthesis and processing of lectin in developing pea cotyledons. J Biol Chem 258:9550–9552

Hoffmann LM, Donaldson DD (1985) Characterization of two *Phaseolus vulgaris* phytohemagglutinin genes closely linked on the chromosome. EMBO Journal 4:883–889

Hoffmann LM, Ma Y, Barker RF (1982) Molecular cloning of *Phaseolus vulgaris* lectin mRNA and use of cDNA as a probe to estimate lectin transcript levels in various tissues. Nucl Acid Res 10:7819–7828

Hopp TP, Hemperly JJ, Cunningham BA (1982) Amino acid sequence and variant forms of favin, a lectin from *Vicia faba*. J Biol Chem 257:4473–4483

Hořejší V, Kocourek J (1974) Studies on phytohemagglutinins. XVII. Some properties of the anti-H specific phytohemagglutinin of the furze seeds (*Ulex europaeus* L.). Biochim Biophys Acta 336:329–337

Hořejší V, Chaloupecka O, Kocourek J (1978a) Studies on lectins. XLIII. Isolation and characterization of the lectin from restharrow roots (*Ononis hircina* Jacq.). Biochim Biophys Acta 539:287–293

Hořejší V, Haskovec C, Kocourek J (1978b) Studies on lectins. XXXVIII. Isolation and characterization of the lectin from black locust bark (*Robinia pseudoacacia* L.). Biochim Biophys Acta 532:98–104

Horisberger M, Vonlanthen M (1980) Ultrastructural localization of soybean agglutinin on thin sections of *Glycine max* (soybean) var. Altona by the gold method. Histochemistry 65:181–186

Horowitz J (1985) A *Phaseolus* mutation results in a reduced level of lectin mRNA. Mol Gen Genet 198:482–485

Horstmann C, Rudolph A, Schmidt P (1978) Isolation, characterization and subunit structure of a phyto-hemagglutinin from seeds of *Vicia faba* L. Biochem Physiol Pflanzen 173:311–321

Hosselet M, van Driessche E, van Poucke M, Kanarek L (1983) L.: Purification and characterization of an endogenous root lectin from *Pisum sativum* L. In: Bøg-Hansen TC, Spengler GA (eds) Lectins: Biology, Biochemistry, Clinical Biochemistry, de Gruyter, Berlin (W) Vol 3, pp 549–558

Hosselet M, van Driessche E, van Poucke M, Kanarek L (1985) The occurence of lectin during the life-cycle of *Pisum sativum* L. In: Bøg-Hansen TC, Breborowicz J (eds) Lectins: Biology, Biochemistry, Clinical Biochemistry, de Gruyter, Berlin (W), Vol 4, pp 583–590

Howard IK, Sage HJ, Stein MD, Young NM, Leon MA, Dyckes DF (1971) Studies on a phytohemaggluti-nin from the lentil. J Biol Chem 246:1590–1595

Howard IK, Sage HJ, Horton CB (1972) Studies on the appearance and location of hemagglutinins from a common lentil during the life-cycle of the plant. Arch Biochem Biophys 149:323–326

Howard J, Kindinger J, Shannon LM (1979) Conservation of antigenic determinants among different seed lectins. Arch Biochem Biophys 192:457–465

Hrabak EM, Urbano MR, Dazzo FB (1981) Growth-phase-dependent immunodeterminants of *Rhizobium trifolii* lipopolysaccharide which bind trifoliin A, a white clover lectin. J Bacteriol 148:697–711

Hwang DL, Yang WK, Foard DE, Lin KTD (1978) Rapid release of protease inhibitors from soybeans: im-munochemical quantification and parallels with lectins. Plant Physiol 61:30–34

Iglesias JL, Lis H, Sharon N (1982) Purification and properties of a D-galactose/N-acetyl-D-galacto-samine-specific lectin from *Erythrina cristagalli*. Eur J Biochem 123:247–252

Irimura T, Osawa T (1972) Studies on a hemagglutinin from *Bauhinia purpurea alba* seeds. Arch Biochem Biophys 151:475–482

Iyer PNS, Wilkinson KD, Goldstein IJ (1976) An N-acetyl-D-glucosamine-binding lectin from *Bandeiraea simplicifolia* seeds. Arch Biochem Biophys 177:330–333

Jaffe CL, Ehrlich-Rogozinski S, Lis H, Sharon N (1977) Transition metal requirements of soybean aggluti-nin. FEBS Lett 82:191–196

Jirgensons B (1978) Circular dichroism studies on the effects of sodium dodecyl sulfate on the conforma-tion of some phytohemagglutinins. Biochim Biophys Acta 536:205–211

Jirgensons B (1980a) Circular dichroism study on structural reorganization of lectins by sodium dodecyl sulfate. Biochim Biophys Acta 623:69–76

Jirgensons B (1980b) Circular dichroism tests on the effect of alkali on conformation of lectins. Biochim Biophys Acta 625:193–201

Kaladas PM, Kabat EA, Kimura A, Ersson B (1981) The specificity of the combining site of the lectin from *Vicia villosa* seeds which reacts with cytotoxic T-lymphoblasts. Mol Immunol 18:969–977

Kalb AJ (1968) The separation of three L-fucose-binding proteins of Lotus tetragonolobus. Biochim Biophys Acta 168:532–536

Kalb AJ, Levitzki A (1968) Metal-binding sites of concanavalin A and their role in the binding of α-methyl-D-glucopyranoside. Biochem J 109:669–672

Kalb AJ, Lustig A (1968) The molecular weight of concanavalin A. Biochim Biophys Acta 168:366–367

Kauss H, Bowles DL (1976) Some properties of carbohydrate-binding proteins (lectins) solubilized from cell walls of *Phaseolus aureus*. Planta 130:169–174

Kauss H, Glaser C (1974) Carbohydrate-binding proteins from plant cell walls and their possible involve-ment in extension growth. FEBS Lett 45:304–307

Kay CM (1970) The presence of β-structure in concanavalin A. FEBS Lett 9:78–80

Kijne JW, van der Schaal IAM, de Vries GE (1980) Pea lectins and the recognition of *Rhizobium legumino-rum*. Plant Sci Lett 18:65–74

Kijne JW, van der Schaal IAM, Diaz CL, van Iren F (1983) Mannose-specific lectins and the recognition of pea roots by *Rhizobium leguminosarum*. In: Bøg-Hansen TC, Spengler GA (eds) Lectins: Biology, Bio-chemistry, Clinical Biochemistry, de Gruyter, Berlin (W), Vol 3, pp 521–529

Köhle H, Kauss H (1979) Binding of *Ricinus communis* agglutinin to the mitochondrial inner membrane as an artefact during preparation. Biochem J 184:721–723

126

Kolberg J, Michaelsen TE (1979) Mitogenic stimulation of human lymphocyte subpopulations by *Lathyrus odoratus* lectin. Acta Path Microbiol Scand Sect C 87:275–279

Kolberg J, Sletten K (1982) Purification and properties of a mitogenic lectin from *Lathyrus sativus* seeds. Biochim Biophys Acta 704:26–30

Kolberg J, Michaelsen TE, Sletten K (1980) Subunit structure and N-terminal sequences of the *Lathyrus odoratus* lectin. FEBS Lett 117:281–283

Kolberg J, Michaelsen TE, Sletten K (1983) Properties of a lectin purified from the seeds of *Cicer arietinum*. Hoppe-Seyler's Z Physiol Chem 364:655–664

Konami Y, Tsuji T, Matsumoto I, Osawa T (1981) Purification and characterization of a Cytisus-type *Ulex europeus* hemagglutinin II by affinity chromatography. Hoppe-Seyler's Z Physiol Chem 362:983–989

Kornfeld K, Reitman ML, Kornfeld R (1981) The carbohydrate-binding specificity of pea and lentil lectins. Fucose is an important determinant. J Biol Chem 256:6633–6640

Kouchalakos RN, Hapner KD (1984) Carbohydrate specificity, metal content and molecular stability of a lectin from sainfoin (*Onobrychis viciifolia*, Scop.). Biochim Biophys Acta 787:237–243

Kouchalakos RN, Bates OJ, Bradshaw RA, Hapner KD (1984) Lectin from sainfoin (*Onobrychis viciifolia* Scop.). Complete amino acid sequence. Biochemistry 23:1824–1830

Kurokawa T, Tsuda M, Sugino Y (1976) Purification and characterization of a lectin from *Wistaria floribunda* seeds. J Biol Chem 251:5686–5693

Lamb JE, Bookstein FL, Goldstein IJ, Newton LE (1981) *Bandeiraea simplicifolia* I isolectins reveal a developmental sequential relationship. J Biol Chem 256:5874–5878

Lamb JE, Shibata S, Goldstein IJ (1983) Purification and characterization of *Griffonia simplicifolia* leaf lectins. Plant Physiol 71:879–887

Lauwereys M, van Driessche E, Strosberg AD, Dejaegere R, Kanarek L (1983) The α- and β-subunits of pea lectin are the result of a post-translational cleavage of a precursor chain. In: Bøg-Hansen TC, Spengler GA (eds) Lectins: Biology, Biochemistry, Clinical Biochemistry, de Gruyter, Berlin (W), Vol 3, pp 603–610

Law IJ, Strijdom BW (1984a) Properties of lectins in the root and seed of *Lotononis bainesii*. Plant Physiol 74:773–778

Law IJ, Strijdom BW (1984b) Role of lectins in the specific recognition of *Rhizobium* by *Lotononis bainesii*. Plant Physiol 74:779–785

Leavitt RD, Felstedt RL, Bachur NR (1977) Biological and biochemical properties of *Phaseolus vulgaris* isolectins. J Biol Chem 252:2961–2966

Liener IE (1986) Nutritional significance of lectins in the diet. In: Liener IE, Sharon N, Goldstein IJ (eds) The Lectins: Properties, Functions and Applications in Biology and Medicine, Academic Press, London, pp 527–552

Lis H, Sharon N (1978) Soybean agglutinin – A plant glycoprotein. J Biol Chem 253:3468–3476

Lis H, Sharon N (1981) Proteins and Nuclid Acids. In: Marcus A (ed) The Biochemistry of Plants. A Comprehensive Treatise. Academic Press, London, Vol 6, pp 371–447

Lis H, Sharon N (1986) Application of lectins. In: Liener IE, Sharon N, Goldstein IJ (eds) The Lectins: Properties, Functions and Applications in Biology and Medicine, Academic Press, London, pp 294–370

Lis H, Fridman C, Sharon N, Katchalski E (1966a) Multiple hemagglutinins in soybean. Arch Biochem Biophys 117:301–309

Lis H, Sharon N, Katchalski E (1966b) Soybean hemagglutinin, a plant glycoprotein. I. Isolation of a glycopeptide. J Biol Chem 241:684–689

Lönnerdal B, Borrebaeck CAK, Etzler ME, Ersson B (1983) Dependence on cations for the binding activity of lectins as determined by affinity electrophoresis. Biochem Biophys Res Comm 115:1069–1074

Lönngren J, Goldstein IJ, Zand R (1976) Circular dichroism studies on the α-D-galactopyranosyl-binding lectin isolated from the seeds of *Bandeiraea simplicifolia*. Biochemistry 15:436–440

Lotan R, Lis H, Sharon N (1975a) Aggregation and fragmentation of soybean agglutinin. Biochem Biophys Res Comm 62:144–150

Lotan R, Skutelsky E, Danon D, Sharon N (1975b) The purification, composition and specificity of the anti-T lectin from peanut (*Arachis hypogaea*). J Biol Chem 250:8518–8523

Manen JF, Pusztai A (1982) Immunocytochemical localisation of lectins in cells of *Phaseolus vulgaris* L. seeds. Planta 155:328–334

Marik T, Entlicher G, Kocourek J (1974) Studies on phytohemagglutinins. XVI. Subunit structure of the pea isophytohemagglutinins. Biochim Biophys Acta 336:53–61

Matsumoto I, Osawa T (1969) Purification and characterization of an anti-H(0) phytohemagglutinin of *Ulex europeus*. Biochim Biophys Acta 194:180–189

Matsumoto I, Osawa T (1970) Purification and characterization of a Cytisus-type anti-H(0) phytohemagglutinin from *Ulex europeus* seeds. Arch Biochem Biophys 140:484–491

Matsumoto I, Osawa T (1972) The specific purification of various carbohydrate-binding hemagglutinins. Biochem Biophys Res Comm 46:1810–1815

Matsumoto I, Osawa T (1974) Specific purification of eel serum and *Cytisus sessilifolius* anti-H hemagglutinins by affinity chromatography and their binding to human erythrocytes. Biochemistry 13:582–588

McCubbin WD, Oikawa K, Kay CM (1971) Circular dichroism studies on concanavalin A. Biochem Biophys Res Comm 43:666–674

McKenzie G, Sawyer WH (1973) The binding properties of dimeric and tetrameric concanavalin A. Binding of ligands to noninteracting macromolecular acceptors. J Biol Chem 248:549–556

Miller JB, Noyes C, Heinrikson R, Kingdon HS, Yachnin S (1973) Phytohemagglutinin mitogenic proteins. Structural evidence for a family of isomitogenic proteins. J Exper Med 138:939–951

Miller JB, Hsu R, Heinrikson R, Yachnin S (1975) Extensive homology between the subunits of the phytohemagglutinin mitogenic proteins derived from *Phaseolus vulgaris*. Proc Natl Acad Sci USA 72:1388–1391

Miller RL (1983) Purification of peanut (Arachis hypogaea) agglutinin isolectins by chromatofocusing. Anal Biochem 131:438–446

Monsigny M, Jeune-Chung KH, Perrodon Y (1978) Separation and biological properties of *Phaseolus vulgaris* isolectins Biochimie 60:1315–1322

Moreira RA, Barros ACH, Stewart JC, Pusztai A (1983) Isolation and characterization of a lectin from the seeds of *Dioclea grandiflora* (Mart.). Planta 158:63–69

Mort AJ, Bauer WG (1980) Composition of the capsular and extra-cellular polysaccharide of *Rhizobium japonicum*. Plant Physiol 66:158–163

Murphy LA, Goldstein IJ (1977) Five α-D-galactopyranosyl-binding isolectins from *Bandeiraea simplicifolia* seeds. J Biol Chem 252:4739–4742

Murphy LA, Goldstein IJ (1979) Physical-chemical characterization and carbohydrate-binding activity of the A and B subunits of the *Bandeiraea simplicifolia* I isolectins. Biochemistry 18:4999–5005

Namen AE, Hapner KD (1979) The glycosyl moiety of lectin from sainfoin (*Onobrychis viciifolia*, Scop.). Biochim Biophys Acta 580:198–209

Neurohr KJ, Young NM, Mantsch HH (1980) Determination of the carbohydrate-binding properties of peanut agglutinin by ultraviolet difference spectroscopy. J Biol Chem 255:9205–9209

Nissen MS, Magnuson JA (1986) Metal-ion binding to tetrameric lima bean lectin. J Biol Chem 261:2514 to 2519

Olsnes S, Saltvedt E, Pihl A (1974) Isolation and comparison of galactose-binding lectins from Abrus precatorius and *Ricinus communis*. J Biol Chem 249:803–810

Olson MOJ, Liener IE (1967) Some physical and chemical properties of concanavalin A, the phytohemagglutinin of the jack bean. Biochemistry 6:105–111

Orf JH, Hymowitz T, Pull SP, Pueppke SG (1978) Inheritance of a soybean seed lectin. Crop Science 18:899–900

Osawa T, Matsumoto I (1972) Gorse (Ulex europeus) phytohemagglutinins. Meth Enzymol 28:323–327

Pandolfino ER, Magnuson JA (1980) $Mn^{2+}$ and $Ca^{2+}$ binding to the lima bean lectins. J Biol Chem 255:870–873

Paulova M, Entlicher G, Ticha M, Koštiř JV, Kocourek J (1971a) Studies on phytohemagglutinins. VII. Effect of $Mn^{2+}$ and $Ca^{2+}$ on hemagglutination and polysaccharide precipitation by phytohemagglutinin of *Pisum sativum* L. Biochim Biophys Acta 237:513–518

Paulova M, Ticha M, Entlicher G, Koštiř JV, Kocourek (1971b) Studies on phytohemagglutinins. IX. Metal content and activity of the hemagglutinin from the lentil (*Lens esculenta* Moench). Biochim Biophys Acta 252:388–395

Père M, Bourillon R, Jirgensons B (1975) Circular dichroism and conformational transition of *Dolichos biflorus* and *Robinia pseudoacacia* lectins. Biochim Biophys Acta 393:31–36

128

Pereira MEA, Kabat EA (1974) Specificity of purified hemagglutinin (lectin) from *Lotus tetragonolobus*. Biochemistry 13:3184–3192

Pereira MEA, Gruezo F, Kabat EA (1979) Purification and characterization of lectin II from *Ulex europeus* seeds and an immunochemical study of its combining site. Arch Biochem Biophys 194:511–525

Pflumm MN, Wang JL, Edelman GM (1971) Conformational changes in concanavalin A. J Biol Chem 246:4369–4370

Poretz RD, Riss H, Timberlake JW, Chien S (1974) Purification and properties of the hemagglutinin from *Sophora japonica* seeds. Biochemistry 13:250–256

Prigent MJ, Bourillon R (1976) Purification and characterization of a lectin (plant hemagglutinin) with N blood-group specificity from *Vicia graminea* seeds. Biochim Biophys Acta 420:112–121

Prigent MJ, Bourillon R (1980) Subunit structure of *Vicia graminea* anti-(blood-group N) lectin. Biochem J 189:185–188

Pueppke SG, Bauer WD, Keegstra K, Ferguson AL (1978) Role of lectins in plant-microorganism interactions. II. Distribution of soybean lectin in tissues of *Glycine max* (L.) Merr. Plant Physiol 61:779–784

Pueppke SG, Friedman HP, Su LC (1981) Examination of Le and lele genotypes of *Glycine max* (L.) Merr. for membrane-bound and buffer-soluble soybean lectin. Plant Physiol 68:905–909

Pull SP, Pueppke SG, Hymowitz T, Orf JH (1978) Soybean lines lacking the 120,000-dalton seed lectin. Science 200:1277–1279

Pusztai AJ (1986) The biological effects of lectins in the diet of animals and man. In: Bøg-Hansen TC, van Driessche E (eds) Lectins: Biology, Biochemistry, Clinical Biochemistry, de Gruyter, Berlin (W), Vol 5, pp 317–327

Pusztai A, Watt WB (1974) Isolectins of *Phaseolus vulgaris*. A comprehensive study of fractionation. Biochim Biophys Acta 365:57–71

Pusztai A, Croy RRD, Stewart JC, Watt WB (1979) Protein body membranes of *Phaseolus vulgaris* L. cotyledons: isolation and preliminary characterization of constituent proteins. New Phytol 83:371–378

Pusztai A, Grant G, Stewart JC (1981) A new type of *Phaseolus vulgaris* (cv. Pinto III) seed lectin: isolation and characterization. Biochim Biophys Acta 671:146–154

Pusztai A, Grant G, Stewart JC (1982) The isolation and characterization of an unusual seed lectin from a "Lectin-free" cultivar of *Phaseolus vulgaris*, Pinto III, and its relationship to lectins synthesized by root cells. In: Bøg-Hansen TC (ed) Lectins: Biology, Biochemistry, Clinical Biochemistry, de Gruyter, Berlin (W), Vol 2, pp 743–758

Reeke GN, Becker JW, Edelman GM (1975) The covalent and three-dimensional structure of concanavalin A. IV. Atomic coordinates, hydrogen bonding and quaternary structure. J Biol Chem 250:1525–1547

Reeke GN, Becker JW, Edelman GM (1978) Changes in the three-dimensional structure of concanavalin A upon demetallization. Proc Natl Acad Sci USA 75:2286–2290

Richardson C, Behnke WD, Freisheim JH, Blumenthal KM (1978) The complete amino acid sequence of the α-subunit of pea lectin, *Pisum sativum*. Biochim Biophys Acta 537:310–319

Richardson M, Campos FDAP, Moreira RA, Ainouz IL, Begbie R, Watt WB, Pusztai A (1984a) The complete amino acid sequence of the major α-subunit of the lectin from the seeds of *Dioclea grandiflora* (Mart). Eur J Biochem 144:101–111

Richardson M, Rougé P, Sousa-Cavada B, Yarwood A (1984b) The amino acid sequences of the $\alpha_1$- and $\alpha_2$-subunits of the isolectins from seeds of *Lathyrus ochrus* (L) DC. FEBS Lett 175:76–81

Roberts DM, Etzler ME (1984) Development and distribution of a lectin from the stems and leaves of *Dolichos biflorus*. Plant Physiol 76:879–884

Roberts DD, Goldstein IJ (1984) Reexamination of the carbohydrate-binding stoichiometry of lima bean lectin. Arch Biochem Biophys 230:316–320

Roberts DM, Walker J, Etzler ME (1982) A structural comparison of the subunits of the *Dolichos biflorus* seed lectin by peptide mapping and carboxyl terminal amino acid analysis. Arch Biochem Biophys 218:213–219

Roth J (1986) Light and electron microscopic localization of cellular glycoconjugates with lectin-gold complexes. In: Bog-Hansen TC, van Driessche E (eds) Lectins: Biology, Biochemistry, Clinical Biochemistry, de Gruyter, Berlin (W), Vol 5, pp 419–431

Rougé P, Chabert P (1983) Purification and properties of a lectin from *Lathyrus tingitanus* seeds. FEBS Lett 157:257–260

129

Rougé P, Labroue L (1977) Sur le rôle des phytohémagglutinines dans la fixation spécifique des souches compatibles de *Rhizobium leguminosarum* sur le pois. C R Acad Sci Paris 284:2423–2426

Rougé P, Plantefol ML (1974a) Etude de la phytohémagglutinine des graines de Lentille au cours de la germination et des premiers stades du développement de la plante. Evolution dans les cotylédons. C R Acad Sc Paris, t. 278:449–452

Rougé P, Plantefol ML (1974b) Etude de la phytohémagglutinine des graines de Lentille au cours de la germination et des premiers stades du développement de la plante. Evolution dans les racines, les teges et les feuilles. C R Acad Sc Paris, t. 278:3083–3086

Roy J, Som S, Sen A (1976) Isolation, purification and some properties of a lectin and abrin from *Abrus precatorius* Linn. Arch Biochem Biophys 174:359–361

Rüdiger H (1977) Purification and properties of blood-group-specific lectins from Vicia cracca. Eur J Biochem 72:317–322

Rüdiger H (1984) On the physiological role of plant lectins. Bioscience 34:95–99

Rutherford WM, Dick WE, Cavins JF, Dombrink-Kurtzman MA, Slodki ME (1986) Isolation and characterization of a soybean lectin having 4-0-methylglucuronic acid specificity. Biochemistry 25:952–958

Schurz J, Rüdiger H (1985) The lectin-binding protein from the pea *(Pisum sativum)*; properties and interactions. Biol Chem Hoppe-Seyler 366:367–373

Sherwood JE, Truchet GL, Dazzo FB (1984) Effect of nitrate supply on the in vivo synthesis and distribution of trifoliin A, a *Rhizobium trifolii*-binding lectin, in Trifolium repens seedlings. Planta 162:540–547

Shoham M, Kalb AJ, Pecht I (1973) Specificity of metal-ion interaction with concanavalin A. Biochemistry 12:1914–1917

Shoham M, Sussman JL, Yonath A, Moult J, Traub W, Kalb (Gilboa) AJ (1978) The effect of binding of metal-ions on the three-dimensional structure of demetallized concanavalin A. FEBS Lett 95:52–56

Sing VO, Schroth MN (1977) Bacteria – plant cell surface interactions: active immobilization of saprophytic bacteria in plant leaves. Science 197:759–761

Sletten K, Kolberg J, Michaelsen TE (1983) The amino acid sequence of the α-subunit of a mitogenic lectin from seeds of *Lathyrus odoratus*. FEBS Lett 156:253–256

Smets G, van Driessche E, Beeckmans S (1985) Ultrastructural localization of pea lectin in the embryo and cotyledons during development by the colloidal-gold method. In: Bøg-Hansen TC, Breborowicz J (eds) Lectins: Biology, Biochemistry, Clinical Biochemistry, de Gruyter, Berlin (W), Vol 4, pp 453–462

So LL, Goldstein IJ (1968) Protein-carbohydrate interaction. XX. On the number of combining sites on concanavalin, the phytohemagglutinin of the jack bean. Biochim Biophys Acta 165:398–404

Solheim B (1983) Purification and characterization of lectins from *Vicia hirsuta*. Physiol Plantarum 58:515–522

Stacey G, Paau AS, Bull WJ (1980) Host recognition in the *Rhizobium* – soybean symbiosis. Plant Physiol 66:609–614

Stahlhut RW, Hymowitz T, Orf JH (1981). Screening of USDA *Glycine soja* collection for presence or absence of a seed lectin. Crop Science 21:110–112

Su LC, Pueppke SG, Friedman HP (1980) Lectins and the soybean-*Rhizobium* symbiosis. I. Immunological investigations of soybean lines, the seeds of which have been reported to lack the 120,000-dalton soybean lectin. Biochim Biophys Acta 629:292–304

Sugii S, Kabat EA (1980) Immunochemical specificity of the combining site of *Wistaria floribunda* hemagglutinin. Biochemistry 19:1192–1199

Sutoh K, Rosenfeld L, Lee YC (1977) Isolation of peanut lectin by affinity chromatography on polyacrylamide-entrapped guar beads and polyacrylamide (Co-allyl α-D-galactopyranoside). Anal Biochem 79:329–337

Talbot CF, Etzler ME (1978a) Development and distribution of *Dolichos biflorus* lectin as measured by radioimmunoassay. Plant Physiol 61:847–850

Talbot CF, Etzler ME (1978b) Isolation and characterization of a protein from leaves and stems of *Dolichos biflorus* that cross reacts with antibodies to the seed lectin. Biochemistry 17:1474–1479

Terao T, Irimura T, Osawa T (1975) Purification and characterization of a hemagglutinin from *Arachis hypogaea*. Hoppe-Seyler's Z Physiol Chem 356:1685–1692

Timberlake JW, Wong RBC, Poretz RD (1980) Properties and subunit characterization of affinity purified *Sophora japonica* lectin. Prep Biochem 10:173–190

130

Tollefsen SE, Kornfeld R (1983) Isolation and characterization of lectins from *Vicia villosa*. Two distinct carbohydrate-binding activities are present in seed extracts. J Biol Chem 258:5165–5171

Toyoshima S, Osawa T (1975) Lectins from Wistaria floribunda seeds and their effect on membrane fluidity of human peripheral lymphocytes. J Biol Chem 250:1655–1660

Trowbridge IS (1974) Isolation and chemical characterization of a mitogenic lectin from *Pisum sativum*. J Biol Chem 249:6004–6012

Tsien HC, Jack MA, Schmidt EL, Wold F (1983) Lectin in five soybean cultivars previously considered to be lectin-negative. Planta 158:128–133

Tsivion Y, Sharon N (1981) Lipid-mediated hemagglutination and its relevance to lectin-mediated agglutination. Biochim Biophys Acta 642:336–344

Uy R, Wold F (1977) 1.4-Butanediol diglycidyl ether coupling of carbohydrates to Sepharose: affinity adsorbents for lectins and glycosidases. Anal Biochem 81:98–107

Van der Wilden W, Herman EM, Chrispeels MJ (1980) Protein bodies of mung bean cotyledons as autophagic organelles. Proc Natl Acad Sci USA 77:428–432

Van Driessche E, Foriers A, Strosberg AD, Kanarek L (1976a) N-terminal sequences of the α- and β-subunits of the lectin from the garden pea *(Pisum sativum)*. FEBS Lett 71:220–222

Van Driessche E, Strosberg AD, Kanarek L (1976b) Studies on the structure of lectins: I. Pea *(Pisum sativum)* lectin. Arch Int Physiol Biochim 84:677–679

Van Driessche E, Vandenbranden S, Kanarek L (1978) Improvement in the purification procedure of pea lectin, and considerations on the subunit structure. Arch Int Physiol Biochim 86:963–964

Van Driessche E, Vandenbranden S, Dejaegere R, Kanarek L (1980) Studies on the subunit structure of *Vicia sativa* lectin. Arch Int Physiol Biochim 88:B50–B51

Van Driessche E, Smets G, Dejaegere R, Kanarek L (1981) The isolation, further characterization and localization of pea seed lectin *(Pisum sativum* L.). In: Bøg-Hansen (ed) Lectins: Biology, Biochemistry, Clinical Biochemistry, de Gruyter, Berlin (W), Vol 2, pp 729–741

Van Driessche E, Smets G, Dejaegere R, Kanarek L (1982) The immuno-histochemical localization of lectin in pea seeds *(Pisum sativum* L.). Planta 153:287–296

Van Driessche E, Lauwereys M, Beeckmans S, Dejaegere R, Kanarek L (1983) Isolation and partial characterization of the pea-lectin precursor. Arch Int Physiol Biochim 91:B42–B43

Van Wauwe JP, Loontiens FG, de Bruyne CK (1975) Carbohydrate-binding specificity of the lectin from the pea *(Pisum sativum)*. Biochim Biophys Acta 379:456–461

Vitale A, Chrispeels MJ (1984) Transient N-acetylglucosamine in the biosynthesis of phytohemagglutinin: attachment in the Golgi apparatus and removal in protein bodies. J Cell Biol 99:133–140

Vitale A, Ceriotti A, Bollini R, Chrispeels MJ (1984a) Biosynthesis and processing of phytohemagglutinin in developing bean cotyledons. Eur J Biochem 141:97–104

Vitale A, Warner TG, Chrispeels MJ (1984b) Phaseolus vulgaris phytohemagglutinin contains high-mannose and modified oligosaccharide chains. Planta 160:256–263

Vodkin LO, Rhodes PR, Goldberg RB (1983) A lectin gene insertion has the structural features of a transposable element Cell 34:1023–1031

Voelker T, Staswick P, Tague B, Chrispeels MJ (1986) The derived amino acid sequence of the seed lectin present in the Pinto Ul 111 cultivar of *Phaseolus vulgaris* and a comparison with PHA-E and PHA-L. In: Bøg-Hansen TC, van Driessche E (eds) Lectins: Biology, Biochemistry, Clinical Biochemistry, de Gruyter, Berlin (W), Vol 5, pp 171–176

Walter P, Blobel G (1981) Translocation of proteins across the endoplasmatic reticulum. III. Signal recognition protein (SRP) causes signal sequence-dependent and site-specific arrest of chain elongation that is released by microsomal membranes. J Cell Biol 91:557–561

Wang JL, Cunningham BA, Edelman GM (1971) Unusual fragments in the subunit structure of concanavalin A. Proc Natl Acad Sci USA 68:1130–1134

Wang JL, Cunningham BA, Waxdal MJ, Edelman GM (1975) The covalent and three-dimensional structure of concanavalin A. I. Amino acid sequence of cyanogen bromide fragments $F_1$ and $F_2$. J Biol Chem 250:1490–1502

Wantyghem J, Goulut C, Frénoy JP, Turpin E, Goussault Y (1986) Purification and characterization of *Robinia pseudoacacia* seed lectins. Biochem J 237:483–489

131

Waxdal MJ (1974) Isolation, characterization and biological activities of five mitogens from pokeweed. Biochemistry 13:3671–3677

Weber E, Neumann D (1980) Protein bodies, storage organelles in plant seeds. Biochem Physiol Pflanzen 175:279–306

Weber E, Manteuffel R, Neumann D (1978) Isolation and characterization of protein bodies of *Vicia faba* seeds. Biochem Physiol Pflanzen 172:597–614

Wèber E, Manteuffel R, Jakubek M, Neumann D (1981) Comparative studies on protein bodies and storage proteins of *Pisum sativum* L. and *Vicia faba* L. Biochem Physiol Pflanzen 176:342–356

Wei CH, Koh C, Pfuderer P, Einstein JR (1975) Purification, properties and crystallographic data for a principal nontoxic lectin from seeds of *Abrus precatorius*. J Biol Chem 250:4790–4795

Wood C, Kabat EA, Murphy LA, Goldstein IJ (1979) Immunochemical studies of the combining sites of the two isolectins A₄ and B₄, isolated from *Bandeiraea simplicifolia*. Arch Biochem Biophys 198:1–11

Wu AM, Kabat EA, Gruezo FG, Poretz RD (1981) Immunochemical studies on the reactivities and combining sites of the D-galactopyranose- and 2-acetamido-2-deoxy-D-galactopyranose-specific lectin purified from *Sophora japonica* seeds. Arch Biochem Biophys 209:191–203

Yamashita K, Hitoi A, Kobata A (1983) Structural determinants of *Phaseolus vulgaris* erythrocytoagglutinating lectin for oligosaccharides. J Biol Chem 258:14753–14755

Yariv J, Kalb AJ, Katchalski E (1967) Isolation of an L-fucose-binding protein from *Lotus tetragonolobus* seed. Nature (London) 215:890–891

Yariv J, Kalb AJ, Levitzki A (1968) The interaction of concanavalin A with methyl-α-D-glucopyranoside. Biochim Biophys Acta 165:303–305

Young NM (1974) The effects of maleylation on the properties of concanavalin A. Biochim Biophys Acta 336:46–52

Young NM, Williams RE, Roy C, Yaguchi M (1982) Structural comparison of the lectin from sainfoin *(O. viciifolia)* with concanavalin A and other D-mannose-specific lectins. Can J Biochem 60:933–941

Young NM, Watson DC, Williams RE (1984) Structural differences between two lectins from *Cytisus scoparius*, both specific for D-galactose and N-acetyl-D-galactosamine. Biochem J 222:41–48

Young NM, Watson DC, Williams RE (1985) Lectins and legume chemotaxonomy. Characterization of the N-acetyl-D-galactosamine-specific lectin of *Bauhinia purpurea*. FEBS Lett 182:403–406

# Additions in Print

While the manuscript was under editorial consideration several most important papers, related to the topics covered by this review, have been published and will be reported in this addendum. Several groups have paid much effort on the elucidation of the posttranslational processing of the Con A precursor to yield the mature Con A subunit as well as its derived fragments (Chrispeels et al. 1986; Bowles et al. 1986; Fay and Chrispeels 1987). By metabolic labeling of immature jackbean cotyledons and analyses of the immunoprecipitated lectin forms, Bowles et al. (1986) proposed the following events to occur: Con A is synthesized as a glycosylated precursor of 33,500 MW which, within a few hours, is processed to a 31,600 MW polypeptide chain. This processing step removes the glycosyl residues from the initial precursor. A proteolytic cleavage within a surface loop of the molecule generates two polypeptides with molecular weigths of 18,800 and 15,100. This cleavage converts the inactive precursor into a molecule with carbohydrate-binding activity.

All further proteolytic cleavages are brought about by a single endopeptidase which specifically recognizes the peptide-bond COOH-terminal to asparagine residues. This enzyme tailors the 15,100 and 18,800 MW peptides to respectively 14,200 and 17,800 MW polypeptides and a covalent bond between the two chains is formed at positions 118 and 119 of the mature protein. From this intermediate, designated as immature Con A, a NH₂-terminal tetra-peptide is finally removed by the endopeptidase that cleaves at the COOH-terminal of Asn to yield the mature Con A chain of 30,000 MW. Consideration of the tertiary structure of the glycosylated precursor and the mature lectin showed that neither of the proteolytic cleavages involved in the processing steps induce major conformational changes.

Although according to Carrington et al. (1985) posttranslational transposition and ligation within the initially synthesized polypeptide is involved in the maturation process of Con A, Bowles et al. (1986) suggest that a transpeptidation event is involved. In this view, the presence of natural fragments in Con A preparations should result from the continued activity of the asparagine endopeptidase, rather than from the persistance of unligated intermediates. The high-mannose oligosaccharide of pro-Con A has been shown to be essential for intracellular transport. Indeed, when glycosylation of the Con A precursor is prevented by tunicamycin, there is little transport of pro-Con A out of the endoplasmatic reticulum to the protein bodies (Faye and Chrispeels 1987). Similarly, it was shown by Bowles et al. (1976) that incubation of cotyledons in the carboxylic ionophore monensin, which, besides other effects, also interferes with terminal glycosylation in the Golgi complex and transport in and out the dictyosomes (Tartakoff 1983), inhibits the conversion of the 33,500 MW polypeptide to any other molecular species. Using immunogold labeling with anti-Con A Bowles and co-workers (1986) could show that, after monensin treatment, electron dense immunoreactive material accumulated between the plasma membrane and the cell wall.

The implication of root lectins in the establishment of legume-Rhizobium symbiosis remains a subject of debate. As mentioned earlier in this paper, the finding that some legume cultivars lack or synthesize only highly reduced levels of seed lectin made some workers suppose that lectins are not involved in symbiosis. Vodkin and Raikel (1986) could show that the roots of Le⁻ soybean varieties, which are devoid of lectin in seeds and roots, contain an antigenically related protein of 33,000 MW. Although it could not yet be determined whether this cross-reactive material displays lectin activity, its localization in vesiculate structures in the root epidermis might indicate that this protein, rather than SBL, is involved in *Rhizobium* recognition.

Using indirect immunofluorescence, Diaz et al. (1986) localized pea root lectin in the surface of four- and five-days old pea roots. Lectin, or a cross-reactive material, was observed on the tips of newly formed growing root hairs as well as on epidermal cells just below the young root hairs. The same pattern was found on super-nodulating as well as on non-nodulating pea cultivars. Similarly, growth of roots in nitrate concentrations which are known to inhibit nodule formation did not affect the distribution of root lectin. Although spot-inoculation tests revealed that susceptibility to *Rhizobium leguminosarum* infection is confined to those regions of the root where lectins are found, it seems thus that other factors rather than, or in addition to root lectins are implicated in successful nodulation. Besides it should be noticed that, in the experimental conditions used, no distinction can be made between lectins and lectin-like cross-reacting materials.

In a recent paper, the role of host root lectins in the specific attachment of *Rhizobium* to legume roots has been questioned again (Smit et al. 1986). These authors reported that pea lectin inhibiting sugars did not prevent the attachment of *Rhizobium leguminosarum* to root hairs. Furthermore, they found that fast-growing heterologous Rhizobia attach nearly equally well to pea root hair tips as *Rhizobium leguminosarum* cells do. The results of Smit and co-workers rather point to the involvement of Rhizobial surface components in host-specific attachment. Indeed, they showed that *Rhizobium leguminosarum* produces extracellular fibrils which resemble fimbriae of Enterobacteraceae. Since a strong correlation was found between the occurence of fimbrillated bacteria and attachment to root hairs it was hypothesized that these fibrils, rather than root surface lectins act as recognition signals between Rhizobia and their respective hosts. Since the fibrils have not yet been purified and characterized, their adhesive properties have not been fully established. However, in view of the fact that the fimbriae of pathogenic bacteria have now clearly been shown to mediate the specific attachment to host tissues (Beachy 1981; Gaastra and de Graaf 1982; de Graaf 1986), it is without any doubt worthwile to look for such structures in both symbiotic and parasitic bacteria which colonize plant tissues. During the last months, several papers have dealt with the further molecular characterization of legume lectins. The lectin from the garden pea *(Pisum sativum)* has been crystallized and its three-dimensional structure was determined at 3 Å-resolution by X-ray diffraction studies (Einspahr et al. 1986). These studies clearly indicate the strong similarity of the secondary and tertiary structure of Con A and pea lectin. Besides it was shown that the amino acid residues surrounding that metal-binding site are identical in both lectins.

Using cDNA cloning and in vitro synthesis of the Dolichos biflorus seed lectin, Schnell et al. (1987) could demonstrate that the two structurally related subunits of the lectin are derived from a single polypeptide precursor by posttranslational processing.

Previously, Vitale et al. (1984b) had shown that the lectin from the cotyledons of *Phaseolus vulgaris* is a glycoprotein which has one high-mannose and one complex oligosaccharide side chain. This finding raised the question why some oligosaccharide chains are processed to form modified chains, while others remain in the high-mannose form. The answer to this exciting question was given by Faye et al. (1986) who showed that the high-mannose chains present on mature glycoproteins remain as such, because they are not sufficiently accessible to the α-mannosidases in the Golgi complex which trims the originally synthesized $Glc_3Man_9(GlcNAc)_2$ precursor to a $Man_5(GlcNAc)_2$ chain, which can then further be modified.

# References

Beachy EH (1981) Bacterial adherence: adhesin-receptor interactions mediating the attachment of bacteria to mucosal surfaces. J Infect Diseases 143:325–345

Bowles DJ, Marcus SE, Pappin DJC, Findlay JBC, Eliopoulos E, Maycox PR, Burgess J (1986) Posttransitional processing of concanavalin A precursors in Jackbean cotyledons. J Cell Biol 102:1284–1297

Chrispeels MJ, Hartl PM, Sturm A, Faye L (1986) Characterization of the endoplasmic reticulum-associated precursor of concanavalin A. J Biol Chem 261:10021–10024

De Graaf FK (1986) The fimbrial lectins of *Escherichia coli*. In: Bøg-Hansen TC, van Driessche E (eds) Lectins: Biology, Biochemistry, Clinical Biochemistry. de Gruyter, Berlin (W), vol 5, pp 285–296

Diaz CL, van Spronsen PC, Bakhuizen R, Logman GJJ, Lugtenberg EJJ, Kijne JW (1986) Correlation between infection by *Rhizobium leguminosarum* and lectin on the surface of *Pisum sativum* L. roots. Planta 168:350–359

Einspahr H, Parks EH, Sugana K, Subramanian E, Suddath FL (1986) The crystal structure of pea lectin at 3.0 Å-resolution. J Biol Chem 261:16518–16527

Faye L, Chrispeels MJ (1987) Transport and processing of the glycosylated precursor of concanavalin A in Jackbean. Planta 170:217–224

Faye L, Johnson KD, Chrispeels MJ (1986) Oligosaccharide side chains of glycoproteins that remain in the high-mannose form are not accessible to glycosidases..Plant Physiol 81:206–211

Gaastra W, de Graaf FK (1982) Host-specific fimbrial adhesins of noninvasive enterotoxigenic *Escherichia coli* strains. Microbiol Rev 46:129–161

Schnell DJ, Alexander DC, Williams BG, Etzler ME (1987) cDNA cloning and in vitro synthesis of the *Dolichos biflorus* seed lectin. Eur J Biochem 167:227–231

Smit G, Kijne JW, Lugtenberg BJJ (1986) Correlation between extracellular fibrils and attachment of *Rhizobium leguminosarum* to pea root hair tips. J Bacteriol 168:821–827

Tartakoff AM (1983) Perturbation of vesicular traffic with the carboxylic ionophore monensin. Cell 32:1026–1028

Vodkin LO, Raikhel NV (1986) Soybean lectin and related proteins in seeds and roots of Le+ and Le− soybean varieties. Plant Physiol 81:558–565

# 4 Illustrations of Lectin-producing Plants (I)
Christa Beurton, Renate Israel, Hartmut Franz

It is the intention of this chapter to convey an idea of the plants which produce lectins and, in so doing are responsible for this part of the biochemist's material. The most important morphological data are given, to which now and then some anecdotal information is added. Those readers, however, who are less interested in these technicalities and wish to gain some visual familiarity may just look at the illustrations.

The present chapter deals with 24 plants species belonging to very different kinds of flowering plants (angiosperms).

More than half of these species belong to the pea family, which is well in keeping with the general prominence of this family in this respect. Although lectin-producing species may be found in approximately 80 families of flowering plants and are also known from algae and lichens, as well as animals, and from roughly 100 species of mushrooms, most of them come from the legumes. The Leguminosae of Fabaceae are one of the largest families among the flowering plants (surpassed only by the Compositae and the Orchidaceae). They include 590–650 genera and 13,200–18,000 species. All of them possess a characteristic fruit, so-called true pod or legume. Most species have root tubercles containing bacteria capable of producing nitrates by the incorporation of atmospheric nitrogen. The family is split into three subunits: the mimosa subfamily (Mimosoideae), the senna subfamily (Caesalpinioideae) and the pea subfamily (Faboideae or Papilionoideae). Frequently, these three subfamilies are treated as distinct families. Most of the lectin producers belong to the Faboideae, and only this subfamily has a pea-like or pea-shaped flower, a corolla of five unequal petals; one outermost (standard or banner), two lateral and similar ones (the wings), and two innermost ventrally united petals, the so-called keel, which enclose the ten stamens and the pistil (see also *Sophora japonica*). In the senna subfamily, the corolla is still irregular (petals are not alike and are not arranged in a radially symmetrical pattern), stamens are ten or fewer, and in the mimosa subfamily the flowers are regular and the stamens are often numerous.

The other ten species belong to seven different smaller or larger families of the flowering plants, some of which are lesser and some better known. It may be added that the majority of the lectins were isolated from seeds, with the exceptions of *Phytolacca americans*, *Solanum tuberosum* and *Robinia pseudoacacia* where they were from roots, tubers and barks, respectively. Lectins of *Viscum album* were found mainly in the leaves and in the stems.

# Abrus precatórius L.

Family:        Leguminosae (Fabaceae), Pea family; Ger. Schmetterlings-
blütengewächse

Common names: Rosary pea, Lucky bean, Crab's eyes, Indian licorice;
Ger. Paternostererbse

Synonym:       *Glycine abrus* L.

The small tropical genus *Abrus* has four to six species, all are shrubs or shrublets, often climbing. Rosary pea is a woody twiner with up to 5 m long branches. The leaves are composed of 8 to 15 pairs of small leaflets which are oblong but with broad ends, and a pointed tip. The pea-shaped flowers are red or purple (seldom white) and form dense inflorenscences with a long stalk. The pods are nearly 4 cm long, oblong and flat. They contain conspicuous, 6–7 mm long seeds. The color of the seeds is scarlet and in the lower third shiny black.

Rosary pea is grown for ornament, and the seeds are frequently used for ornamental jewelry, such as necklaces and rosaries. The Latin epithet precator means "one who prays".

In India, they are also used as weights (rati). The seeds are highly poisonous inside, the toxic substance is abrin, a protein. When chewed thoroughly, a single seed may be deadly. The roots of this plant may serve as a licorice substitute (hence the name Indian licorice). In East Africa, the seeds are said to possess magic powers. Another species serves man as antidote when stung by scorpions.

Rosary pea is native to the Old World tropics. Undoubtedly, the seeds were taken to the New World in times of the slavetrade. Today, this species is widely naturalized in the tropics. The genus' name is Greek and means delicate, referring to the small leaflets.

R. Israel

*Abrus precatorius*

# Arachis hypogaéa L.

Family:            Leguminosae (Fabaceae), Pea family; Schmetterlings-
blütengewächse
Common name:   Peanut; Ger. Erdnuß

The peanut is probably native to South America, and there it has been cultivated in gardens for thousands of years. In North America, the plant was still unknown in the eighteenth century. In other subtropical regions cultivation of this plant is steadily expanding: 10%–15% starch, 25%–35% protein, and 42%–48% fat in the seed make it an ideal nutritive.

Peanuts are annual, lightly hairy bushy herbs up to 60 cm tall or they may develop spreading or running characteristics. The leaves have two opposite pairs of oval or elliptical leaflets and two small leaf-like organs at the leaf base. The pea-shaped flowers are about 2 cm in height and a long tube which surrounds the ovary and the style conveys the impression of the flowers being supported by a long stem. The development of the peanut's fruit is a botanical curiosity. While the golden-yellow (rarely white) flowers are visible on the stems, the fruit none the less ripens in the soil (hence the Latin epithet "hypogaea"). This is accomplished like this: After self-pollination (rarely cross-pollination by insects) and fertilization of the ovules a specific tissue from below the ovary rapidly elongates, thus forming what is called the peg which possibly gains a length of 15–20 cm. Soon the peg penetrates the soil with the ovary and the dried remnants of the flower on its tip. There the ovary ripens into the pod (or "nut") which contains one to three edible seeds. All this is presumably a protective device against drought.

The tetraploid *Arachis hypogaea* with 2n = 40 chromosomes is a result of the combination of sets of chromosomes from two or more wild species. More than 10 wild species of the genus *Arachis* are known, mostly from Brazil, and most of them are diploid with 2n = 20 chromosomes. A remarkable difference between wild peanuts and the cultivated species is the morphology of the fruits: The pods of cultivated peanuts are only one-chambered. The seeds ripen simultaneously, and the whole plant is easily pulled out, which is, of course, significant for industrial production.

The derivation of the name *Arachis* is obscure. Plinus has used this name for plants with subterranean fruits.

R. Israel

*Arachis hypogaea*

# Canavália ensifórmis (L.) DC.

Family:          Leguminosae (Fabaceae), Pea family; Ger. Schmetterlings-
                 blütengewächse
Common names: Jack bean, Horse bean; Ger. Jackbohne, Riesenbohne, Schwertbohne
Synonym:         *Dolichos ensiformis* L.

The genus *Canavalis* is widely distributed in the tropics and represented by different species in each hemisphere. All species have large and woody sword-shaped pods (12–30 cm long) and large seeds. The genus' name is aboriginal. The Latin epithet "ensiformis" stands for sword-like or sword-shaped.

Jack bean is native to tropical America (Mexico to Brazil and Peru, West Indies). Seeds of this species from 2000 years ago were found in Mexico, Arizona and Peru. In South America this plant is characteristic in gardens of the native population. The erect or semi-erect or slightly climbing bushes are usually cultivated as annuals and 1–2 m in size. The large and long stalked leaves are composed of three elliptic or ovate-elliptic or oblong leaflets, 6–20 cm in length. The large pea-shaped flowers are purple or red-violet and they grow in dense inflorescences which are 5–12 cm long and pendulous. The pods are quite conspicuous (size in Fig. reduced), 20 to 30 cm long and 2–5 cm broad. A single fruit contains 12–18 white and compressed seeds, up to 2 cm in length. Jack bean is cultivated throughout the tropics for green manure, as a soil cover against erosion, and for forage. Young fruits and seeds yet unripe may be eaten as a vegetable. Ripe beans are mildly poisonous and edible only when thoroughly cooked.

Jack bean is closely related to *Canavalia brasiliensis* MART. ex BENTH., the latter possibly forming the original stock. There is an opinion (Brücher 1977) that the Jack bean presents a transition stage toward domestication.

*Canavalia ensiformis*

R. Israel

141

# Datúra stramónium L.

Family:            Solanaceae, Nightshade family; Ger. Nachtschattengewächse
Common names: Thorn apple, Jimson weed; Ger. Stechapfel, Teufelsapfel

The nightshade family is a large family with 85 genera and about 2300 species in the tropics as well as the temperate regions. Many of them are highly poisonous. They are used for ornament, for food and drugs. *Datura stramonium* is probably native to subtropical North America and to South America but is widely distributed throughout the tropics and subtropics. It has been naturalized in temperate regions. It occurs as weed in fields and waste lands.

Thorn apple is an annual characteristically branched plant, growing to a height of 1.20 m with ovate and lobed leaves up to 20 cm long. The flowers have a white or violet color and a trumpet-shaped corolla up to 10 cm long. They bloom from June to October. The fruit is usually a prickly or spiny ovoid capsule up to 5 cm in length. There are also varieties with a smooth surface. The capsules contain numerous black and flat seeds.

Various species of the genus occurred in folklore, religion and were used in medicine long before the thorn apple came to Europe. Also today, *Datura* is extensively used for its narcotic and hypnotic properties. Thorn apple is one of the most poisonous plants. In the Middle Ages, it was grown only for ornament in Europe, and first in Spain. Then it was made use of for medical purposes. Like Henbane *(Hyoscyamus niger)* and Belladonna *(Atropa belladonna)*, the thorn apple was used as an ingredient of love potion. Since the seventeenth century it has gone wild in Europe.

Today it is cultivated in the United States and Europe for the drug stramonium. The active principles are alkaloids, including hyoscyamine, atropine and scopolamine. Stramonium is known for its narcotic effects and it is also good against asthma (it relaxes the bronchial muscles). The leaves serve for cigarettes for asthmatics. The genus' name **Datura** is a vernacular East-Indian name.

142

R. Israel

Datura
stramonium

143

# Euónymus europaéus L.

Family:            Celastraceae, Staff-tree family; Ger. Baumwürgergewächse
Common name:  Spindle tree; Ger. Europäisches Pfaffenhütchen

The spindle tree is a shrub or small tree up to 3 m tall with opposite-standing, ovate or oblong-lanceolate leaves. Young branches are distinctly four-angled. The small and yellowish-green flowers are inconspicuous like the weak inflorescences which are composed of three to seven flowers. They blossom in May and June. The fruits are ripe by October. They form a deeply four-lobed red or pink capsule up to 2 cm across. The fruit shape is reminiscent of a Catholic priest's barret, hence the German name "Pfaffenhütchen". The seeds are white and enclosed in an orange cover (aril). When the ripe fruit bursts the seeds remain attached to filaments hanging out of the fruit. In this state, the orange aril forms a beautiful contrast to the fruit color. The seeds are eaten and distributed by birds. Like all of the more than 120 species belonging to this genus, the spindle tree is poisonous in all its parts.

The spindle tree ist native in temperate Europe and Western Asia. Seeds of this plant are known since the tertiary. It grows in the bushes as well as in humus-rich woods of deciduous trees. In autumn, the leaves turn wonderfully reddish. Because of its attractive appearance in autumn it is grown in gardens. *Euonymus* is an ancient Greek name. It may be also spelled *Evonymus*.

The spindle tree is a member of the staff-tree family to which belong trees and shrubs (often climbing) of about 60 genera, i.e., more than 850 species. They are widely distributed all over the world. All members of this family have alternate or opposite and undivided leaves and small regular flowers which are mostly bisexual. They consist of four of five sepals and petals, four or five or rarely ten stamens and one pistil. The fruits are of different types (e.g., capsules or berries) but all contain seeds with a brightly colored, conspicuous and fleshy aril.

# Euonymus europaeus

R. Israel

# Húra crépitans L.

Family:          Euphorbiaceae, Spurge family; Ger. Wolfsmilchgewächse
Common name:   Sandbox tree; Ger. Sandbüchsenbaum

*Hura crepitans* is a tree up to 25 m high with very thorny trunk and branches. The species grows in the tropical coastal regions, it is native to the region as the West Indies to Bolivia. Because of its large heart-shaped leaves it is often grown as a shade tree in the tropics. Each tree carries two types of flowers. The male flowers are deep red and stalkless; they arise from an elongate main axis and form terminal spikes; female flowers are solitary and much greater. They consist of a green pistil only. The ovary is not three-chambered like most in the spurge family, but with more than 10 (up to 25) cavities. When the fruits ripen, the walls between the cavities become hard and woody. The ripened capsule has a diameter of about 8 cm. When the fruit dries, each of the cavity walls expands from triangular to U-shape and the fruit explodes with a loud report (hence the name in the West Indies "monkeys dinner bell" or "árbol del diablo" and the Latin epithet "crepitans" for exploding). The seeds are shot out, up to 15 m away, the greatest distance reached by any member of this family.

The latex of this species is very poisonous and it is said to cause blindness. It is used for drugging fish in the Amazon region. The seeds and their oil are used as a purgative.

Before the blotting-paper era the fruits were used as sandboxes. The unripe capsules were tied together with wire, the seeds extracted and replaced by sand.

R. Israel

*Hura crepitans*

# Labúrnum alpinum (MILL.) BERCHT. et J.S.PRESL

Family:          Leguminosae (Fabaceae), Pea family; Ger. Schmetterlings-
                 blütengewächse
Common name: Scotch laburnum; Ger. Alpen-Goldregen
Synonym:         Cýtisus alpinus MILL.

The name *Laburnum* is of ancient Latin origin. Frequently, the species of this genus are included among the brooms (i. e., genus *Cytisus*). Scotch laburnum is a native of Southern Middle-Europe, Italy and West-Balkan. It is used as an ornamental, as is golden chain *(Laburnum anagyroides)*. Scotch laburnum differs from golden chain in that it grows stiffer and more upright. Usually, it is a tree of 3−4 m in height, and sometimes it reaches up to 10 m. The leaves are on long stalks and divided into three elliptic leaflets. In June and July, when in blossom, a wonderful odor arises from the yellow pea-shaped flowers. This and its large and pendulous inflorescences lend this plant a very attractive appearance. The pods are compressed and contain several seeds. In contrast to golden chain, the pods are nearly without hairs and their upper margin winged.

All parts, but especially the seeds (up to 3 %) and leaves (0.3 %) of Scotch laburnum and golden chain contain poisonous alkaloids.

Laburnum
alpinum

R. Israel

149

# Láthyrus odorátus L.

Family:          Leguminosae (Fabaceae), Pea family; Ger. Schmetterlings-
                 blütengewächse
Common name:  Sweet pea; Ger. Duftplatterbse, Duftwicke, Gartenwicke

The genus *Lathyrus* with about 160 annual or perennial herbs is widely distributed in the Northern hemisphere. It is closely allied taxonomically to *Vicia* (vetch) and *Pisum* (pea). The sweet pea is an annual tall-climbing herb with slightly hairy and winged stems. There are also nonclimbing varieties (dwarf sweet pea) such as the kind known as cupid. The leaves consist of one pair of oval or elliptical leaflets and a terminal branched tendril. The long stalked flower cluster consists of a few one to three conspicuous and pea-shaped flowers. There are numerous color variants, for instance violet, purple, pink, blue, white brown and yellow, or flowers with red standard and white wings. Varieties with very large flowers are called Spencer types. The pods of the sweet pea are nearly 5–6 cm long and hairy. They contain in most cases several ball-like seeds.

Sweet pea is a native of South Italy and Sicily, however it is grown the world over as an ornamental plant because of its attractive and agreeably smelling flowers (hence the Latin epithet "odoratus"). The plant covers garden fences, pergolas and arcades, cottage walls and balconies. Flowering time ranges from June to autumn.

The first known record is from the Dutch botanist Commelin in *Hortus Medicus Amsterodamensis* in 1697/1701, who got the seeds from Pater F. Cupani. At the beginning of the eighteenth century the sweet pea was kept in some gardens in London, and later on in the gardens of Western Europe.

The flowers of the sweet pea are used for production of essential oils.

*Lathyrus odoratus*

R. Israel

151

# Láthyrus satívus L.

Family:          Leguminosae (Fabaceae), Pea family; Ger. Schmetterlings-
                 blütengewächse
Common names: Grass pea, Chickling pea, Chickling vetch;
                 Ger. Saatplatterbse, Deutsche Kichererbse


Grass pea is an annual suberect or climbing herb with winged and much-branched stems. The leaf stalks are long and winged as well. The leaves are divided into one pair of leaflets and a long branched tendril. The leaflets are linear-lanceolate (5–7.5 cm long and 3–7 mm wide) and they have parallel veins thus being somewhat similar to the grasses in appearance (hence the English name grass pea). At leaf base there are two long and leaflet-like organs. The pea-shaped flowers are 1.2–2 cm long, mostly solitary and long stalked. Size and color of the flower leaves are variable, and mostly the petals are white with blue veins, or the standard and wings are pink or bluish. The pods are nearly oblong, 2.5–3.8 cm long, broad and flat, and dorsally two-winged. They contain three to five characteristically angular and wedge-shaped seeds, somewhat smaller than peas, and of different color (e.g., white, grayish-brown, yellowish), usually spotted or mottled.

The grass pea is an old cultivated plant in the Mediterranean region, Western and Central Asia, in the Southern and Southeastern parts of Europe (partly in Central Europe), in India, Iran and Western Siberia. Its origins are unknown. Probably it is native to Western Asia, and the Mediterranean region is a secondary center of diversification. The grass pea is grown for food and forage. Leaves and young pods are eaten as vegetable. Seeds are used mainly as a forage crop, but they also serve for human consumption, particularly in the form of meal. The seeds, especially the dark colored ones, contain unusual and poisonous amino acids (not participating in protein formation). Consumption may or may not cause a sickness called "lathyrismus". Today, only India is an important producer of this plant.

R. Israel

Lathyrus
sativus

# Lens culináris MEDIK.

Family:            Leguminosae (Fabaceae), Pea family; Ger. Schmetterlings-
                   blütengewächse
Common names: Lentil, Gram; Ger. Linse
Synonyms:          *Ervum lens* L., *Lens esculenta* MOENCH

There are about half a dozen species of the genus *Lens* in the Mediterranean and in Western Asia. One species, the lentil, is cultivated for its nutritious seeds. It was known to the people of ancient Egypt and ancient Greece, i.e., around 6000 B.C. There are two hypotheses about the lentil's origin: Probably it came from the Near East and Mediterranean region, or this is a secondary center and their primary center of diversity is in Southwest Asia (Afghanistan to the Himalayas). Today, lentils have been introduced and cultivated in most subtropical and warm temperate regions of the world, and higher altitudes of the tropics in both hemispheres.

The seeds of lentil are generally familiar, but not so the plants. It is an annual much branched, bushy herb, softly hairy and with slender stems, 25–75 cm tall. The leaves consist of four to seven pairs of oval, 1–2-cm long leaflets ending in a tendril or bristle; stipules are small. The flowers are small and inconspicuous, mostly of pale blue color. They are clustered in a small one-to-three flowered inflorescence with a slender and weak stalk. The pods are compressed, smooth, oblong, up to 1.5 cm long and nearly as broad. They contain one or two seeds the shape of which is just that of a lens. The seeds may vary in color from green or greenish-brown to light red speckled with black. There are many varieties which differ in color of the flowers, and in color, shape and mottling of the seeds.

The earliest seeds known as used by man are very small, approximately 2 mm across. Lentil seeds from the Bronze Age already have a diameter of 3–4 mm. Today, there are varieties with seeds of 8 mm in diameter, prefered, however, only in Europe.

R. Israel

*Lens*
*culinaris*

# Lycopérsicon esculéntum MILL.

Family:          Solanaceae, Nightshade family; Ger. Nachtschattengewächse
Common name:  Tomato; Ger. Tomate
Synonym:      *Solánum lycopérsicum* L.

The tomato is a branched, rough-hairy and glandular herb of strong smell. It is erect or trailing and reaches up to 1.5 m in height. Though it is a perennial, it is mostly grown as an annual. Blossoms are yellow, and the fleshy fruit is a true berry with many seeds. The tomato's characteristics are very variable from kind to kind but they are also strongly environmentally conditioned. Probable different varieties are contributed to the one or two hundred kinds grown. The varieties differ in size, shape and texture of fruits; these being for instance either small and globular or oblong, or larger and pea-shaped or globular and more than 7.5 cm across. Fruit color is mostly red, seldom yellow, or pink.

The tomato was cultivated in Mexico in pre-Columbian times, and Mexico was the center of its domestication. It comes probably from South America (Peru, Ecuador) which is inhabited by approximately half a dozen wild tomato species. The Spanish explorers introduced it to Southern Europe. Only very much later it was introduced to Northern Europe and North America, and then only for ornament because it was considered to be poisonous (hence probably the Greek genus' name Lycopersicon, "wolf peach"). Today the tomato is one of the vegetables with great economic importance. It ranks next to potatoes and sweet potatoes. The fruits are very rich in vitamins. They are eaten raw or cooked, and ketchup and juice may be made of them. Green tomatoes may be made into pickles.

*Lycopersicon esculentum*

R. Israel

# Momórdica charántia L.

Family:            Cucurbitaceae, Gourd family; Ger. Kürbisgewächse
Common names: Balsam pear, Bitter gourd; Ger. Balsambirne, Bittergurke

*Momordica charantia* is a slender but high-climbing annual with more or less hairy stems. The leaves are rather thin, and deeply lobed. Like in *Hura crepitans* there are male and female flowers on the same individual. The petals are yellow or yellow-orange, and at the base or at least not above the middle of the peduncle (the stalk of the inflorescence) there is a prominent leaflet. The fruits are oblong or oval, becoming slimmer toward the ends, about 10–20 cm in length, and covered with blunt warts. The orange-yellow color of the fruit contrasts exotically with the fleshy scarlet of the seed covers (aril), which become visible when the fruit bursts at maturity. The seeds are about 12 mm long, gray or brown with prominent patterns.

Balsam pear may be native to the Old World tropics. Like other species of the genus, we know of about 60, it is widely cultivated throughout the tropics for its flowers, but principally for its bitter but edible fruits. The balsam pear and the balsam apple *(Momordica balsamina)* are the species of this genus the young fruits of which turned into pickles, curries and salads.

Other popular species of the gourd family, grown for edible fruits, for ornament, the production of containers or music instruments, and as vegetable sponges are *Citrullus lanatus* (watermelon), *Cucumis sativus* (cucumber), *Cucumis melo* (melon), *Cucurbita pepo* (summer squash, winter squash, pumpkin), *Lagenaria siceraria* (bottle gourd) and *Luffa cylindrica* (vegetable sponge).

R. Israel

Momordica
charantia

159

# Phaséolus vulgáris L.

Family:         Leguminosae (Fabaceae), Pea family; Ger. Schmetterlings-
                blütengewächse
Common names: Bean, Common bean, Kidney bean, Field bean; Ger. Gartenbohne

The common bean is a highly variable annual herb. There are erect varieties, 20–60 cm tall (common bush bean), and twining varieties with stems 2–4 cm long (common garden pole bean). The green or purple leaves of this species are divided into three ovate stalked leaflets, having one or two leaf-like organs at the leaflet's base. There are also stipules on the leaf base. The color of the nearly 1 cm long pea-shaped flowers is mostly white, seldom yellow, pink or purple. The pods are 8–20 cm long and 1–1.5 cm wide, and with a prominent beak. They vary in color and shape. There are green-podded and wax-podded varieties. The pods contain 4–6 (–12) seeds, which also vary in shape and color (globular or oblong; white, brown, purple, grey or blue-black; often speckled).

The common bean is an important vegetable plant, young pods and ripe seeds are eaten. It is native to the New World (South America to Mexico). In Peru it was eaten already around 8000 B.C. and 30% of today's world production is in Latin America. The Spaniards and the Portuguese brought the common bean to Europe. In Central Europe it gradually replaced the broad bean *(Vicia faba)* as an important vegetable. Today, the common bean is the most widely cultivated bean in temperate regions, and in the tropics and subtropics as well. More than 500 (up to 1000) varieties are known.

R. Israel

Phaseolus vulgaris

# Phytolácca americána L.

Family: Phytolaccaceae, Pokeweed family; Ger. Kermesbeerengewächse
Common names: Virginian pokeweed, Pokeberry, Poke, Scoke;
Ger. Amerikanische Kermesbeere, Schminkbeere
Synonym: *Phytolacca decandra* L.

*Phytolacca americana* is a member of a small plant family with about 125 species, mostly native in the tropical and subtropical parts of America, some of them occurring in Africa, Madagascar and Asia. Many species of this family are medicinal and narcotic, and others are grown in gardens as ornamental plants or as vegetable.

Virginian pokeweed is a tall herb up to 3.50 m high with fleshy succulent stems. The leaves are 10–30 cm long, oblong or ovatelanceolate. When boiled and specially treated (because of the saponins present) they are edible and reminiscent of spinage. The small bisexual flowers, about 4–6 mm across, are clustered in tall and long-stalked inflorescences. Each flower possesses 5 to 16 free or united ovaries, 6–23 free stamens but only one set of green or somewhat colored floral leaves. The darkish purple and fleshy berries (about 8 mm across) arouse a feeling of exotic exuberance. They contain very small black seeds. In Central Europe pokeweed is grown in the gardens as an ornamental with the fruits as the main element. It is native to temperate North America but is nowadays widely distributed in subtropical and tropical regions. In the Mediterranean the berries are grown for improving the color of red wine and for staining confectionery. The characteristic red pigment in the berries of *Phytolacca americana* L. is betacyanin. It has an essentially identical structure with betanin, the pigment of *Beta vulgaris* (Beetroot).

R. Israel

*Phytolacca americana*

163

# Písum satívum L.

Family:        Leguminosae (Fabaceae), Pea family; Ger. Schmetterlings-
               blütengewächse
Common names: Common pea, Garden pea; Ger. Erbse

The common pea is a smooth and bushy or climbing annual herb, usually bluish-green with tender stems, 30–150 cm long. The leaves consist of one to three pairs of leaflets and a terminal branched tendril; leaflets usually oval, 1.5–6 cm long and approximately 1 cm broad. Conspicuous are two leaf-like organs at the leaf base (stipules), usually they are larger than the leaflets, up to 10 cm long. The flowers are solitary or in two or three flowered inflorescences; petals are white, or pink, or purple. The pods are swollen or compressed, short stalked, straight or curved, 4–10 cm long and 1.5–2.5 cm wide. They contain two to ten ball-like or irregularly shaped seeds, smooth or wrinkled, whitish grey, green or brownish.

The common pea grows the world over. It is one of the oldest cultivated plants and is known already from the Stone Age. Its origins are unclear and no wild type is known. Some believe it comes from ancient Egypt, and others would like it to come from Europe to Western Asia.

There are numerous garden races (e. g., early dwarf pea, edible podded pea, field pea). These forms belong to *Pisum sativum* L. in a broad sense. They are cultivated for the fresh green seeds, green pods, dried seeds and for foliage. The oil from ripe seeds has antisex hormonic effects, and produces sterility and antagonizes the effects of the male hormone.

In the temperate regions garden pea forms the most important human nutriment among the legumes. World output amounts to 10.2 million t per annum.

R. Israel

Pisum
sativum

165

# Rícinus commúnis L.

Family:             Euphorbiaceae, Spurge family; Ger. Wolfsmilchgewächse
Common names: Castor oil plant, Castor bean, Palma Christi;
                    Ger. Wunderbaum, Rizinus

This tropical herb is probably native to North Africa but is widely distributed elsewhere. In the tropics and subtropics it makes a small tree (up to 12 m), in temperate regions it may be grown as a summer annual, reaching 1–4.5 m in height. The leaves are very large and 5–11 lobed in a hand-like fashion, up to 1 m across and with a long leaf stalk. The flowers grow together in a large cluster up to 60 cm long and without petals. Within the cluster the yellow-green male flowers are below, and the greenish female flowers above; flowering commences from the base. The stamens are very numerous and their filaments are much branched. The deep-red upper parts of the female flowers are conspicuous. The fruit is a 1.2–2.5 cm long spiny capsule. It contains large characteristically marked and colored seeds.

The castor oil plant is a variable species with numerous varieties which differ in size, color of stems and leaves, concentration of glands on the leaf stalk, or in size and markings of the seeds. There are varieties with red stems and bluish-green or dark red leaves, or with bright green leaves and white veins. Other forms have capsules without spines.

The species has been cultivated in Egypt since about 4000 B.C. for its remarkably oil-rich seeds. The thick colorless or greenish oil is obtained by pressing. It is used in oil lamps, as an excellent lubricant (especially for airplanes), for medical purposes as a purgative and otherwise. World production of seeds per year is up to 1.8 million t. Important producers are Brazil, Mexico, India, the Soviet Union and some African countries. The seeds of the castor oil plant are highly poisonous. They contain ricin which is one of the most poisonous natural agents.

It inhibits protein synthesis, and then seeds are sufficient to kill a man. None the less, oil may be secured from the seeds by pressing, because the poison remains in the pulp. The castor oil plant is also a very popular ornamental plant. Ricinus is a classical Latin name.

R. Israel

*Ricinus communis*

# Robínia pseudoacácia L.

Family:            Leguminosae (Fabaceae), Pea family; Ger. Schmetterlings-
                   blütengewächse
Common names:  Locust tree, Black locust, False acacia; Ger. Robinie, Falsche Akazie,
                   Scheinakazie


The locust tree is native to eastern North America (from Pennsylvania to Georgia) and widely cultivated in other areas of North America as well as in Europe, where it has gone wild and is firmly established. Originally it was grown for ornament, and later as a forest tree. It is a good pioneer on waste areas, and forms on slopes a protection against erosion. *Robinia pseudoacacia* L. is a tree up to 30 m tall with deeply furrowed dark-brown bark and prickly branches. The leaves are divided into 7–19 thin leaflets, oval or elliptic, of 2.5–4.5 cm in length. The flowers are of a sweet odor and about 2 cm in length, they are clustered in long pendulous, softly hairy inflorescences. The pea-shaped petals are white with the exception of a yellow or yellow-green spot on the banner. The flowers blossom in May and June. The pods are linear-oblong, reddish-brown, 5–10 cm long, and smooth; they remain on the branches during winter. They contain some brown or black-brown, and 5–6 mm long seeds.

There are many varieties. The flower colors may differ, some varieties have unarmed branches or pendulous ones etc. The wood of *Robinia pseudoacacia* L. is extremely heavy, it is resistent to decay, and its durability is three times that of oak.

The first *Robinia* from North America was introduced in France by Vespasian Robin (a herbalist) about 1620.

Robinia
pseudoacacia

R. Israel

# Solánum tuberósum L.

Family:          Solanaceae, Nightshade family; Ger. Nachtschattengewächse
Common name:  Potato; Ger. Kartoffel, Erdapfel

The potato is an erect and branching, more or less spreading perennial herb 0.40–1.20 m in height. Usually it is grown as an annual. Its leaves are composed of three or four pairs of leaflets and smaller ones in between. A somewhat limited number of bluish, yellowish, whitish or purple flowers with a tubular corolla form together a long stalked and forked inflorescence. The fruit is a small inedible berry of varying color (yellowish, green or brownish-green, or purple). The roots are fine, fibrous. The potatoes, however, are no product of the roots, but underground stem-tubers.

The usual way of propagation is vegetatively by the tubers or parts thereof, the so-called "seed potatoes". The numerous varieties regulary cultivated have been produced through a long history of selection and hybridization. The varieties' adaptive range extends to a broad array of soil conditions and potatoes grow at altitudes up to 2500 m.

The potato is an American native. Several wild species grow from Mexico to Chile. Probably, *Solanum tuberosum* is a native from the Central Andes region (Peru, Bolivia). The first published record of this species is in Pedro Cieza de León's Cronica del Peru from 1553. The Spaniards brought the potato to Spain during the last decades of the sixteenth century. Somewhat later it came to the British Isles. A century later it had spread throughout Europe. Interestingly, this plant came to Central Europe from the British Isles and not from Spain. It came to North America together with Irish immigrants in 1719. Today, the potato is grown the world over (except the tropical regions). Important producers are the Soviet Union, the United States, Poland, the G.D.R., the F.R.G. and France, and about 50% of the potato crop is produced in Europe (including the Soviet Union). The total volume of annual production exceeds that of some cereals e. g., rye *(Secale cereale)*. This, however, is due in part to the 78% of water content of potatoes, which is also more than that of any cereal. The carbohydrates (sugar and starch) have a share of 18% and proteins and fat are also present. The starch occurs in characteristic grains. The mealy texture of the potato when cooked is brought about by the swelling of the grains and the bursting of the cell walls. Potatoes, of course, are essential in human nutrition. They also form a valuable stock feed. They are used for the production of starch as well as alcohol.

R. Israel

*Solanum
tuberosum*

# Sophóra japónica L.

Family:          Leguminosae (Fabaceae), Pea family; Ger. Schmetterlings-
                 blütengewächse
Common names: Japanese pagoda tree, Chinese scholar tree; Ger. Japanischer Schnur-
                 baum, Pagodenbaum

There are about 55 species in the genus *Sophora*, mostly in the warm temperate and subtropical regions of both the hemispheres. All the species are trees with ornamental bark and very hard wood. *Sophora japonica* is a tall handsome tree up to 25 m or more in height. Despite its name it is native to China and Korea. It is remarkable for its wrinkled and contorted trunk and branches. The deciduous leaves are composed of up to 17 separate leaflets. The small creamy white flowers are pea-shaped as in the majority of the pea family. The corolla (the second set of floral leaves, typically brightly colored or white) consists of five petals, one of which is large and erect (standard), the two lateral petals smaller with narrowed basal portions (the wings), and the two lowermost ones united to form the keel. The flowers of the pagoda tree are born in clusters up to 25 cm in length. They appear only on old trees and then in late summer and autumn, and they attract many insects. The fruits themselves are very decorative in shape. The pods are constricted like a few thick pearls on a necklace. They contain one to four black seeds.
The Japanese pagoda tree is vigorous and hardy and grows in gardens of Central Europe as well as in the Mediterranean. The variety "Pendula" has long slender pendulous branches, and "Violacea" has a white standard but wings and keel are tinged purple.
*Sophora* is an Arabic name of a tree with pea-shaped flowers.

Sophora
japonica

R. Israel

173

# Tríticum vulgáre VILL. (= Tríticum aestivum L.)[1]

Family:          Gramineae (Poaceae), Grass family; Ger. Süßgräser
Common name:  Wheat; Ger. Saatweizen

Like other cereals (e. g., rice, maize, oat, barley and rye) the genus *Triticum* is a member of the grass family. There are approximately 8000 species in this family. To man, it is one of the most important plant families, for it supplies him with food, forage, materials for house-building, and various species serve for ornament and so on. The flowers of the grasses are highly specialized and minute. They are assembled into inflorescences. The characteristic fruit of this family is the caryopsis, containing one seed only, and fruit and seed are mostly enseparably linked, and form a hard grain.

The common wheat is grown in innumerable varieties which differ in both morphological and physiological characters. There are truly annual races (the Latin word "aestivum" means "pertaining to summer") which are sown early in the year and are called spring wheat, and there are winter-annual races. Winter or fall wheat is sown in autumn and matures during the following season. There are hard wheats with small and protein-rich grains and soft wheats with larger ones which are richer in starch. Other physiological characters include such features as resistance to cold or drought and time of maturing. The common wheat is the most important bread-wheat and one of the most important of man's food plants. The earliest traces of wheat are from Mesopotamia, Anatolia and from Iran, and are nearly 8000 years old, which is long before Jesus' prayer "Give us this day or daily bread". Wheat was the principal food in settled communities of the Mediterranean, in Europe and temperate Asia, and even today in modern societies it is the most important vegetable food. The hexaploid *Triticum aestivum* is known only cultivated, and its place of origin is unclear. It is a hybrid, probably as a result of the combination of sets of chromosomes from three species genera *(Triticum* and *Aegilops)* through a long history of selection. Common wheat grows northward up to the 68th degree of latitude, is adapted to all moderately dry temperate climates, and is not grown in warm humid regions. Main centers of cultivation are between the 30th and 60th degree of north latitude and between the 25th and 40th degree of south latitude. *Triticum* is the Latin name for wheat.

---

1   The correct name is *Triticum aestivum* L. In lectinology, however, the species is known by its synonym *Triticum vulgare* VILL. Other synonym: Triticum sativum LAM.

Triticum
aestivum

R. Israel

# Ulex europaéus L.

Family:          Leguminosae (Fabaceae), Pea family; Ger. Schmetterlings-
                 blütengewächse
Common names: Furze, Gorse, Whin; Ger. Stechginster, Gaspeldorn

*Ulex* is an ancient Latin name.

*Ulex europaeus* is a dense twiggy, spiny, dark-green bush, 50–150 cm in height. Stems, sepals, stalks of flowers and pods are hairy. Mostly the leaves are reduced to green scales or thorn-like leaf stalks, both being shortly and softly hairy. Only young plants and shoots near the ground bear fully developed leaves with leaflets. The pea-shaped flowers become crowded toward the twigs. Petals are bright yellow and full of scent, the sepals (the outermost set or floral leaves as a whole) are yellow, and two-lipped; lower lip minutely three-toothed, upper minutely two-toothed, teeth curving toward each other. The pods are small (about 15 mm long), oblong, shaggy and dark-brown. The fruits contain each only a few appendaged seeds.

Furce is native to Western and Southern parts of Europe, mainly in rough grassy places and on heaths, usually in light soils. It has been introduced to other parts of Europe and to North America. The flowers blossom in spring and early summer and frequently once more in autumn. In gardens you encounter this species as cover plant and as ornamental.

R. Israel

Ulex
europaeus

# Vícia crácca L.

Family: Leguminosae (Fabaceae), Pea family; Ger. Schmetterlings-
blütengewächse
Common names: Bird vetch, Bird's tare, Cow vetch; Ger. Vogelwicke

Bird vetch is a widely distributed plant in Europe and Asia, which has also settled in North America. It grows mostly in meadows as well as pastures and waste places. It is a perennial climber or trailer with angular stems of up to 1.3 m. The leaves are usually green and composed of 4–12 pairs of linear or linear-oblong leaflets and a terminal tendril. Some varieties possess silky white leaves. The smallish purple and pea-shaped flowers are concentrated into one-sided inflorescences which arise from the leaves' axils. They appear throughout the summer. The withering flowers turn whitish. The small pods contain five to eight black-brown seeds. Bird vetch served as forage, and at times it has been employed for ornament.

There are more than 150 species in the genus, mostly occurring in temperate areas. Only some of them are grown for food, forage, green manure or ornament. The genus' name is a classical Latin name.

*Vicia cracca* L. is a member of the pea family, which comprises some of the most important economic species among the flowering plants. The pods in this family contain more protein material than any other vegetable product. Members of this family are used for food, fodder and oils. Garden peas *(Pisum sativum)*, lentils *(Lens culinaris)*, peanuts *(Arachis hypogaea)*, beans (e.g., *Phaseolus vulgaris*), or soybeans *(Glycine max)* are some of the outstanding food plants, and soybeans and vetches (e.g., *Vicia faba*) for instance, are among the outstanding forage plants. Famous ornamentals in the pea family are for instance wisteria (e.g., *Wisteria floribunda*), sweet pea *(Lathyrus odoratus)*, brooms, senna *(Cassia)*, or orchid tree *(Bauhinia variegata)*.

*Vicia cracca*

R. Israel

179

# Vícia fába L.

Family:          Leguminosae (Fabaceae), Pea family; Ger. Schmetterlings-
                 blütengewächse
Common name:     Broad bean, Faba bean, Horse bean; Ger. Ackerbohne, Saubohne,
                 Pferdebohne, Puffbohne
Synonyms:        *Faba bona* MEDIK., *Faba vulgaris* MOENCH


*Vicia faba* is known only cultivated and the Latin epithet is the classical name for bean. The wild progenitor of this old cultivated plant has not been discovered but several wild species, especially the *Vicia narbonensis*-complex, are close relatives. However, attempts to cross them with *Vicia faba* L. have failed up to date.

The broad bean is a more or less upright annual plant with large and unbranched stems up to 2 m tall. There are also dwarf forms, 30–40 cm in height and with a rich foliage. The bluish-green and fleshy leaves are composed of one to three pairs of mostly alternate, large leaflets, 3 to 8 cm in length. The white pea-shaped flowers, 2–4 cm in lenght, have purple markings on their wings. One to five flowers are clustered on a short stalk in the axils of the leaves. The pods are large and thick (usually 5–12 cm in length and 1–3 cm wide), greenish-black, or brown to black. They contain three or four (five) seeds; shape and color vary from angled to nearly globular and from bright reddish-brown to dark-purple, often mottled or dotted. There are some varieties with taller pods up to 30 cm or more.

The broad bean is an important vegetable because of its fruits and seeds which are rich in proteins. It is one of the most important winter crops in the Middle East for man's consumption. In Central Europe it was an important vegetable up to the sixteenth and seventeenth century before being replaced by other cultivated plants which were introduced from the New World (e.g., beans, potatoes, or maize). Today, it serves mainly as a fodder crop in Central Europe. It is cultivated in nearly all temperate, warm temperate and subtropical countries. Western Asia is probably a primary center of diversification and the Mediterranean region is a secondary one.

Vicia faba

R. Israel

# Víscum álbum L.

Family:          Loranthaceae[1], Mistletoe family; Ger. Mistelgewächse
Common name:  Mistletoe; Ger. Mistel, Europäische Mistel, Weiße Mistel

The mistletoe family consists of herbs or shrubs which are semiparasitic on trees. They extract nourishment from their host with special root-like organs (haustoria). There are about 40 genera and 1400 species, mostly confined to the tropical forests with a few genera like *Viscum* and *Loranthus* extending into the temperate regions. The stemps usually branch in pairs, generally with swollen nodes from which the leaves arise. The undivided leaves are thick and leathery, sometimes reduced to small scales. The leaves are usually in pairs on the opposite sides of an axis or they are whorled.

*Viscum album* has green, much-branched stems up to 1 m and evergreen, yellow-green leaves of 4–8 cm in length. The plants are of exceedingly slow growth. During the winter the ball-like green bushes on the trees are especially conspicuous. *Viscum album* L. grows on deciduous trees only. There are some more mistletoes in Europe. *Loranthus europaeus* is confined on oaks (Ger. Eichenmistel).

The small unisexual flowers of *Viscum album* usually have the sexes on different plants; floral leaves are greenish, sepal-like, usually four; male flowers exceed the female ones in size. The inflorescences consist of 3–5 short-stalked flowers. The flowers bloom in February, March or April.

The berry-like fruit is a product of the ovary and the surrounding axis. It is sticky, white (very seldom yellowish-white), and 5–10 mm across. The fruits are eaten by birds and the seeds are passed out of the alimentary tract thus facilitating migration of the plant among the trees. *Viscum* is ancient Latin and means bird lime; the sticky mass of the berries used to be employed for bird catching.

In Europe, especially Britain, mistletoe serves as a Christmas decoration often together with holly *(Ilex aquifolium)* and ivy *(Hedera helix)*, and their colorful berries. The American counterpart used for this custom is *Phoradendron flavescens.*

---

1  Nowadays, the family Loranthaceae generally is split up into two or more families and the genus *Viscum* is attached to the Viscaceae. However, the Viscaceae are closely related to Loranthaceae and both are probably of common origin.

Viscum
album

♀

♂

R. Israel

# Index

The asteriks (*) behind the numbers indicate the illustrations of lectin-producing plants.

185

186

Pseudomonas putida   115
Pseudomonas tomato   115–116
Psophocarpus tetragonolobus   **49**

# R

radio immunoassay (RIA)   111, 117, 119
Raubitschek, H.   22
Reid, H.F.   21
Reval (now Tallinn, USSR)   13
Rhizobium japonicum   105, 107–108,
   117–118
Rhizobium (leguminosarum)   105–108,
   133
Rhizobium phaseoli   105, 118
Rhizobium trifolii   106
ricin   10, 12–13, 16–24
ricin, dose response   17–19
ricin, effect, agglutination   18, 19, 21–24
ricin, immunity   12, 19, 20
ricin preparation, isolation   10–12, 15–16,
   21
ricin toxicity   13, 16, 18–19, 23–24
Ricinus communis   10, 13, 15–16, 18, 23,
   29, 35, **49–50**, 57, 70, 73, 108, **166**, 167*
Ritthausen, H.   12, 15, 23
robin   21
Robinia pseudoaccacia   **50**, 82, 94, 135,
   **168**, 169*
root lectins   107–108, 112, 133
Rostock   13

# S

Salvia slarae   **70**
Sambucus   **50–51**
Sarothamnus scoparius   **51**
SDS-electrophoresis   100, 107, 113
Secale cereale   **51**
seed lectin   108–112, 115–116, 118
Sesamum indicum   **51**
Solanum tuberosum   **51**, 135, **170**, 171*
Sophora japonica   29, **51**, **70**, 82, 96, 104,
   135, **172**, 173*
Staphylococcus aureus   104
Stillmark, P.H.   10, 13, **15–18**, 20–24, 73
storage proteins   28
Strasbourg   12, 15
subcellular fractionation   109
subunit heterogeneity   101–104

# T

Tetracarpidium conophorum   **70**
Trichosanthes kirilowii   **52**
trifoliin A   106
Trifolium pratense   105–106
Trifolium repens   **52**, 96, 107–108
Triticum vulgare (= T. aestivum)   **52, 174**,
   175*
Tulipa gesneriana   **53, 70–71**
turbidimetry   35

# U

Ulex europaeus   **53**, 69, 82, 95, **176**,
   177*
Urtica dioica   30, **53–54**
UV detection (spectra)   82
UV spectra   87–92
UV spectroscopy   107

# V

Vicia bungei   **54**
Vicia cracca   30, **54–55**, 97, 101, **178**,
   179*
Vicia ervilea   **54**, 97
Vicia faba   6, **54–55**, 82, 97, 105, 108, 114,
   150, 178, **180**, 181*
Vicia graminea   55, 97
Vicia hirsuta   **55**, 97
Vicia narbonensis   180
Vicia sativa   **55**, 82, 97
Vicia tetrasperma   **55–56**
Vicia unijuga   **56**
Vicia villosa   **56**, 97
Vigna radiata   **56**, 96
Vigna unguiculata   **56**
Viscum album   **56–57**, 70, 135, **182**,
   183*

# W

Werner, E.   11–12, 15–16, 21, 23
Wisteria floribunda   29–30, **57**, 94, 178
Woronzow, W.N.   21

# X

Xanthomonas phaseoli   116

# Z

Zea mays   **57**

## Volume 2 – 1989

Lectins as Mitogens • *C. A. K. Borrebaeck, R. Carlsson*
Viscaceae Lectins • *H. Franz*
Mechanism of Action of Ricin and Related Toxic Lectins
on the Inactivation of Eukaryotic Ribosomes • *Y. Endo*
Effects on Gut Structure, Function and Metabolism
of Dietary Lectins. The Nutritional Toxicity of the
Kidney Bean Lectin • *A. Pusztai*
Potential Participation of Tumor Lectins in Cancer Diagnosis,
Therapy and Biology • *H.-J. Gabius*

## Volume 3 – 1990